home
book of
taxidermy
.and
tanning

home book of taxidermy and tanning

GERALD J. GRANTZ

STACKPOLE BOOKS

HOME BOOK OF TAXIDERMY AND TANNING

Copyright © 1969 by Gerald J. Grantz
Published by
STACKPOLE BOOKS
Cameron and Kelker Streets
Harrisburg, Pa. 17105

First printing, October 1969
Second printing, February 1970
Third printing, May 1970
Fourth printing, October 1970
Fifth printing, June 1971
Sixth printing, July 1972
Seventh printing, October 1973
Eighth printing, April 1974
Ninth printing, July 1974
Tenth printing, May 1975
Eleventh printing, January 1976
Twelfth printing, March 1976
Thirteenth printing, August 1976
Fourteenth printing, February 1977
Fifteenth printing, September 1977
Sixteenth printing, February 1978
Seventeenth printing, November 1978
Eighteenth printing, July 1980
Nineteenth printing, August 1981
Twentieth printing, October 1981
Twenty-first printing, May 1982
Twenty-second printing, April 1983
First paperback printing, September 1984
Second paperback printing, May 1985
Third paperback printing, October 1986
Fourth paperback printing, April 1987
Fifth paperback printing, May 1988

Printed in the U.S.A.

Library of Congress Cataloging in Publication Date

Grantz, Gerald J
    Home book of taxidermy and tanning.

    Includes index
    1. Taxidermy.   2. Tanning.   I. Title.
QL63.G7   579'.4   77-85651
ISBN 0-8117-2259-7

Contents

and Head Details • Sewing, Trimming and Adjustments • Finishing Up
the Head • Final Touches
WHOLE BODY MOUNTS
OPEN-MOUTH MOUNTING

MOUNTING ANTLERS AND HORNS
ORDERING LEATHER AND SKIN NOVELTIES
MAKING BIG-GAME FOOT NOVELTIES

PRESERVING SMALL SPECIMENS
BIRD SKINS
SMALL MAMMAL SKINS
HANDLING TRAPPED OR OTHER LIVE SPECIMENS
RECORD KEEPING

IMPORTANCE OF CORRECT PROCESSING
HISTORICAL NOTES
NEEDED TOOLS AND EQUIPMENT
Knives • Currier's Knife • Fleshing Knife • Slicker • Fleshing Beams •
Pulling Bench
SKINNING AND HANDLING
SALTING AND CURING
SOAKING, DEGREASING, AND FLESHING
FORMULAS FOR TANNING SOLUTIONS

TANNING PROCESSES FOR LEATHER AND FURS
Oxalic Acid and Small Skins • Salt, Alum and Fur Skins • Alum-Car-
bolic Acid Soak • Alum Paste and Small Fur Skins • Pickle-Neats-foot
Oil Tanning • Salt-Acid Tanning • Combination Mineral and Vegetable
Tanning for Fur Skins
MAKING RAWHIDE
DRY-CURING SKINS
MAKING BUCKSKIN
Indian Buckskin Making
TANNING SNAKESKINS
CHROME TANNING FOR HEAVY LEATHER
Skinning, Salting, Storing • Fleshing • Dehairing • Tanning • Softening
SKINS AND FURS—CARE AND CLEANING
Deodorizing Skins • Cleaning Furs • Mothproofing

Foreword

Contrary to popular notion, taxidermy and simple forms of tanning can be learned easily by almost anyone. There is nothing especially difficult or mysterious about the work.

The word "taxidermy," stems from the Greek word "taxis," meaning order, and "derma," skin, the word having been defined as "...the art of preparing and preserving the skins of vertebrates and mounting them in a lifelike manner."

Very few hobbies can provide so much enjoyment for such a small investment as taxidermy. It is indeed made to order for those, young or old, who love to hunt and fish and who enjoy field sports. Some of the finest work is possible even when using the simplest tools and materials such as are usually available in the average household.

With the help of this book, an interested person should be able to undertake successfully, spare-time hobby projects of mounting fish or game trophies of his choice. Should the amateur's interest in taxidermy grow and he wish to go on to projects demanding even greater skills, this book will have given much insight regarding the specialty and the necessary foundation on which to build.

This book tells how to mount typical trophies such as squirrels, fox and other small animals, various-sized birds including hawks and pheasants, big-game trophies such as deer and bear, and fish—whether large or small.

It also tells what the beginner needs to know about tanning hides and making buckskin, rawhide, snakeskins, leather, and how to process and prepare fur pelts. It explains, too, how to deodorize furs, and how to clean and mothproof them.

For those who collect specimens for study, the book has suggestions about the care and preservation of skins.

Dealing first with the working area, tools and materials generally employed in taxidermy, the book describes how to make and use papier-mache, plaster of paris molds and paper animal, bird and fish forms, skills basic to doing good taxidermy. Chapters on mounting fish, birds, and animals follow.

The steps in making simple antler and horn mounts, leather, skin and foot novelties are also explained.

For those interested in processing fur pelts, making rawhide, buckskin, tanning snakeskins or making leather there is basic information in the last several chapters which cover also the make-up and use of twelve tanning solutions and detailed explanations on how to apply each of seven tanning processes.

Dr. William T. Hornaday and later, Carl E. Akeley were among the first to raise taxidermy to its present rightful status as an art, with Akeley generally credited with the development of the method of sculpturing clay models of specimens and preparing true-to-life forms on which prepared skins could be mounted. Though better form materials are now readily available, this method is still in use today. James L. Clark followed Akeley and became one of the modern leaders in the field.

Most of the tools used in taxidermy and tanning are properly kept sharp and users are reminded to guard against accidental cuts. It is distressing to nick oneself while skinning out a specimen, and one must always try to avoid such possible paths of infection.

A book such as this does not attempt, of course, to include *all* the fine points of technical information familiar to professionals, nor does it pretend to qualify its readers for embarking on professional work. While it contains much authoritative information—quite enough for the production of high-quality work—the results will depend much on the reader's aptitude, skill and perseverance, and upon his earnestness in trying to do good work. And, for his first several mounting jobs, the reader should not set too high a standard for his work, for one is bound to make a few slip-ups. Proficiency will nevertheless come easily enough as experience is gained on simple projects.

G.J.G.

CHAPTER 1

The Home Taxidermist's Working Area And Equipment

THE FOREMOST CONSIDeration when starting taxidermy is finding a suitable place to do the work. Since a considerable amount of dirt and debris is inevitable, the selected location should be agreed upon with other members of the household. Wives and mothers will usually agree to some reasonable compromise when tactfully approached, and indeed they generally turn out to be of great help. It pays to have them involved from the start.

WORK AREA

Locate the work table or bench in a well lighted place, near a water supply, if possible. There should be ample space for tools and materials within easy reach. These should be laid out in a methodical manner so as to be instantly available when needed. This planning-ahead system is really the best way to get things done. Neatness and order simplifies the work!

Basement, unused room, shed or garage may prove suitable according to individual requirements. Oftentimes, a combination of areas will be used; for example, skinning, fleshing, or form-making operations might be done in the garage or shed, while actual mounting work can be done in more comfortable quarters in the house.

TOOLS

Tool requirements for amateur taxidermy are few and simple, most of these being available in the average household or local community. Dealers in taxidermist supplies provide fast and convenient mail-order service. Taxidermy supply catalogs are a valuable source of information and the amateur should obtain several and familiarize himself with their contents. Knowing in advance what is available for the job at hand can be of great help.

It is really unnecessary and unwise to lay out any appreciable amount of money for specialized tools before one is certain he wishes to continue the work and whether he has an aptitude for it.

Here is a list of suggested tools for a modest beginning and remarks about each:

Knife

The first requisite of any knife is that it be sharp and hold its edge. Select knives of good quality, high carbon steel and choose a size in keeping with the job required of it. A good pocket or penknife with a thin keen blade can be used on birds, fish and small mammal jobs. Single-edged razor blades and craft knives serve many purposes. Better, is a fixed-bladed surgeon's scalpel, used by all taxidermists and ideally suited for the work. These are available at physicians' supply houses and taxidermist mail-order dealers.

Use a paring knife and a broad-bladed butcher's knife for heavier duty. Sheath knives used by hunters serve the purpose provided they have no hilt or guard extending at right angles to the blade. The so-called "blood-groove" on any knife is unnecessary, and serves only to weaken the blade.

Whetstone

Sharp cutting edges are a must in taxidermy and tanning. Some blades hold their edges much longer than others. Obtain an Arkansas stone (composed of pure silica) or a Washita oilstone and use it often.

Skin Scrapers

A curved grapefruit knife with serrated edges is ideal for fish, bird

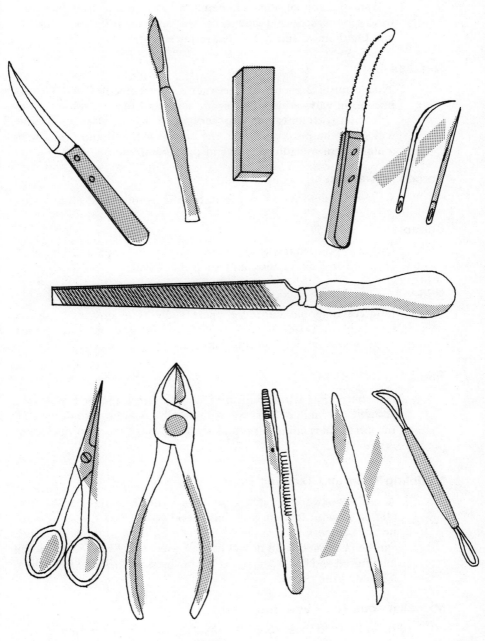

Tools.

and small mammal skins. Homemade scrapers are easily fashioned from spoons, steel shoehorns, paint scrapers and the like. Sharpen and file them to suit individual requirements.

Needles

Taxidermists use curved, 3-corner surgical needles and glovers' needles in various sizes for sewing skins and furs. These are available in assortments and are very inexpensive. Darning needles will serve the purpose. Set in the end of a small stick, they form a handy tool for adjusting eyelids, preening, etc.

Forceps (Tweezers)

Six-inch straight types are useful both in pointed and blunt styles.

Scissors

Ordinary pointed household scissors and small shears can be used. Have several sizes handy and keep them sharp.

Nippers

Pointed, diagonal cutting nippers such as electricians use are invaluable in taxidermy work. They are used to cut wire, bones, cartilage, tough leg tendons, cardboard fiber, etc.

File

For sharpening wires used in mounting. Use a ten or twelve-inch "double-cut" mill or flat type with handle. A motor grinder speeds the job of sharpening wires and serves many other purposes, if one is available.

Modeling Tools and Tamping Rods

A set of boxwood modeling tools in various shapes can be purchased, or whittle them from hardwood dowel rods. Tamping rods, hooks, scrapers, etc., made to suit each purpose as the need arises, can be fashioned from dowel rods or pieces of heavy wire and be hammered and filed to shape at the business end with a loop handle formed at the other.

Miscellaneous Tools and Equipment

In addition to these tools, the taxidermist will find use for most of the following items:

twist drill set

hack or meat saw

pliers

trowel or putty knife

tape measure

small rule reading in millimeters (for measuring eye sizes)

brushes and combs

tinner's shears

mixing bowls and tins

artist's and regular paint brushes

spoons

stapler

assorted boxes, plastic buckets and other containers

MATERIALS FOR GENERAL USE

The following list covers the basic materials required for simple mounting of most birds, mammals and fish. Read the subsequent chapters dealing with specific mounting instructions to determine the individual job material requirements. It is important and again strongly urged that all tools and materials required for the scheduled work be at hand and laid out in an orderly manner before proceeding.

Borax

Powdered borax may be purchased almost anywhere and is inexpensive. It is a preservative and its use has largely replaced methods of poisoning skins by arsenic. Unlike arsenic, borax is absolutely safe to use and does not irritate the skin. Use borax by rubbing it on all skins and bones. Artificial bodies should be moistened slightly with water, then rubbed with borax. To make a borax-water wash for cleaning and soaking skins, sprinkle enough borax in warm water to make a saturated solution; that is, add enough so that some remains undissolved on the bottom of the container after stirring. Borax is also excellent for fluffing and drying skins that have been washed. Work the powder in well, then beat it out with a small stick when it has dried.

Borax Mixes and Alternatives

Some taxidermists feel that valuable mountings must be poisoned to insure their complete long-term protection against spoilage and the ravages of insects. This could be true if the specimens are subjected to humid and dusty conditions, especially if the skins were not thoroughly fleshed and cleansed before mounting.

For those amateurs who feel they must use something more potent than borax in their work it is suggested that equal parts of powdered arsenic (arsenic trioxide) and borax be mixed together in a dry state and applied to the skins and bones of specimens. Under extremely damp and humid conditions and climates, use alum instead of borax in the mixture. Add about one ounce (teaspoonful) of powdered arsenic per gallon of borax-water soaking solutions.

In a sense, arsenic is no more dangerous to use than many common poisons found in the average home. The chief hazard in this (and any other poison) is the possibility of its being used by mistake in place of something else, so make sure the poison container is labeled "Poison" and keep it locked up in a safe place.

No one should employ such noxious preparations without being fully familiar with their handling hazards.

Excelsior

Sometimes called wood wool. Excelsior is used for making artificial body forms and for filling and drying purposes. Because excelsior is widely used for packing fragile articles it is easily obtainable at drug stores or upholstery shops, plate glass dealers, etc., where it can generally be had for the asking. Finer grades of this material, and tow, a softer hemp or sisal material, can be purchased from dealers in scientific supplies or taxidermist supply houses. This material remains a valuable asset to the taxidermist because of its many qualities unavailable in any other form.

Modeling Clay

A permanently pliable clay with an oil base variously named "Plastilina," "Art Clay," etc., sold at art supply counters. The cheap toy store varieties often prove adequate. Potter's clay, or the natural earth clays found in some areas are sometimes used, but these deteriorate in time and are generally considered unsatisfactory for permanent work. Potter's clay is much used but it shouldn't be because of its corrosive nature.

Papier-Mache

Pronounced as "paper mashay." The commercial product is made in dry form. When mixed sparingly with water it forms a pliable modeling material, much like putty, usable for a great many purposes in taxidermy. Papier-mache can be had in either fast or slow-setting varieties, and when hardened, it can be smoothed and worked with edged tools. It can also be colored by painting or by adding pigments to it during the mixing process. Most taxidermists use this material for setting eyes, molding nostrils, making artificial forms, rock stands, fish work, etc. The amateur should order a quantity from his dealer or make his own as needed.

Salt

Use fine common table salt (without special additives) for all work. Salt is cheap; always use plenty.

Cotton

Upholsterers and furniture repair shops usually have the coarse type. Clean, soft grades of batting useful in taxidermy are often available at notions counters and drug stores. It is unnecessary to buy the expensive sterile variety.

Cotton Cord

Used for wrapping bodies of excelsior and in windings to hold plumage in place while drying. It is much cheaper when purchased in 2-1/2 lb. cones.

Thread

Carpet or machine thread, preferably linen, will cover the requirements for most sewing operations. Department stores generally carry these threads in various sizes, or possibly the local tailor or shoemaker can supply it. Use a strong twine for heavy stitching, and lighter linen threads for smaller work and minor repairs.

Beeswax

A small piece is sufficient. Use it to strengthen and preserve sewing threads. Simply rub the sewing cord over the cake before use.

Wire

Use only galvanized, annealed wire, obtainable at hardware stores. Size 8 to 18 (new standard gauge) will cover most jobs. Select a size just large enough to support the mount firmly.

Remember, too, that a mount stiffens up while drying, even though it is a bit wobbly while in a "green" state.

Suggested wire sizes for a few representative birds and mammals are given for comparative purposes:

Starling	— Weasel	— 18 gauge
Crow	— Gray Squirrel	— 14 gauge
Ringneck Pheasant	— Woodchuck	— 12 gauge
Eagle	— Fox	— 8 gauge

Ear Liners

It is best for the amateur to purchase paper liners for his first attempts at game-head mounting. Later, he may wish to make up his own liners from perforated sheet lead, plastic, leather, laminated paper, and the like.

Oil Paints and Brushes

It is necessary to restore colors to all fish after the mounting work is completed. Oil paints in tubes may be purchased at any art store. The following basic colors will cover most work: burnt sienna, burnt umber, viridian green, permanent green (light), lemon yellow, yellow ochre, cadmium yellow (medium), alizarin crimson, cadmium red (light), ultramarine blue, cerulean blue, zinc or titanium white and ivory black. Obtain also a bottle of prepared painting oil for use as a conveyor and mixer.

Four sizes of flat-style bristle brushes, #4, 6, 10 and 12, and a #4 oil color sable brush (pointed) should be enough for the amateur. Detailed suggestions on oil painting will be found later in the book (see Chapter 3).

Shellac

Buy both orange and white shellac in small amounts only as needed, to be assured of freshness. Check dates on the containers.

Sand

Used in one method described for fish mounting. Obtain enough clean sand to fill the largest fish skin anticipated. The important requirements of the sand is that it be free of dirt and thoroughly dry throughout. Some building supply dealers and hardware stores carry air-cleaned play sand in convenient bags. Sand is also used in plaster molding operations.

Plaster of Paris

Use fresh, high-grade molding plaster for fish mounting, form-making operations, etc. Plaster is also valuable for cleaning, drying, and fluffing feathers and fur skins that have been washed. The powder is rubbed in well, then removed with its accumulation of dirt and moisture by shaking and blowing with a hair dryer or the blower attachment of a vacuum cleaner. Available wherever building materials are sold. Do not use "gauging" plaster.

Plastic Foams

Urethane foams, resulting from the mixing of two separate chemical preparations, combine with air, expanding to about thirty times their original volume. The resultant blend dries out into rigid form with a very short interval, often in merely ten to fifteen minutes. The employment of urethane is quite extensive now in industry; we see various blocks of the rigid product quite often, particularly in packing and packaging use. Most often the product is white or uncolored, but it can be given color.

This material is produced by spraying the chemicals into a form (or the chemicals can be poured from a container) and although the chemicals combine with dramatic speed, producing a foam that expands rapidly, the foam expansion is said to exert pressure so gentle that it will not, for instance, force open the cover of a closed cardboard box.

Despite this, it will fill out any container or form quite snugly, forming a substance which, when dry, contains thousands of tiny air cells and is remarkably light, waterproof, and highly buoyant. Urethane foam will adhere to metal, wood, concrete, insulation and most other materials. Sprayed on a water surface, it will float, but continue to expand and harden. It is regarded as impervious to decay from soil organisms and not damaged by chemical actions.

Interesting as all this may seem, this material's use in taxidermy represents so far mainly an area open to experimentation. One might be tempted, perhaps, to experiment with it in mounting a very large game fish such as tarpon where, except through the use of hollow forms, a lightweight mount is not easily achieved.

For those with the inclination to try working with it in taxidermy, the material can be obtained in small spray can containers under the trade-name of Minit-Foam; write to Kerr Chemical,

Inc., 1001 Northwest Highway, Des Plaines, Illinois 60017. For larger quantities, consult dealers specializing in marine materials and paints; merely ask about the availability of plastic foams which can be mixed by the user for various do-it-yourself projects.

Small Items

A number of small items should be gotten together such as brass nails, wood screws, pins, tacks, cardboard, glue, notebook and pencil, to mention a few.

GLASS EYES AND BODY FORMS

These must be purchased as needed.

Glass Eyes

Formerly strictly an European product, quality glass eyes are now produced in America and are obtainable from dealers in taxidermist supplies. It is advisable to have eyes of the proper size and color on hand before any work is started on a specimen. Send for a catalog and become familiar with the sizes, shapes and colors available for the birds and animals likely to be mounted. A suggestion which may solve this problem is to obtain a set of uncolored, crystal eyes in all the graduated sizes. Then, when the specimen to be mounted is at hand, the correctly-sized eyes may be selected and hand-painted on their backs, using an artist's brush and oil colors.

Fish eyes have irregular black pupils with burnt-in colors for the commonly-mounted species. These are also available in clear (flint) style for hand-coloring. Color fish eyes by cementing a small stick to the back of the eye, centered on the pupil for use as a hand-hold; then paint the glass while viewing it from above. Remember that many fish eyes have a gold edging around the black pupil. Use bronze, silver and gold metallic enamels in conjunction with oil paints for lifelike eye effects.

The following comments may be especially of help to readers preparing to order glass eyes. Dealers in taxidermy supplies list suggested sizes and colors for the common birds, animals, and fish in their catalogs, and will usually send suitable artificials if the species are known to them. Manufacturers use the metric system for determining glass eye diameters in millimeters (m/m). Thus, a gray squirrel would take an eye approximately 12 m/m or size 12. Solid black eyes are suitable for some of the rodents such as rats, mice, and squirrels. They are also satisfactory for skunks, mink, otter and others of the weasel family. Raccoons take a black eye. Many animals take a brown eye having a round black pupil. The cat family require a special eye, as do the foxes, deer, fish, etc.

Approximate glass eye sizes for a few of the common animals are:

Mouse	#3
Rat, Mole	#4
Weasel	#6
Mink, Skunk	#8
Porcupine, Badger	#11-12
Squirrel, Groundhog	#11-12
Raccoon, Cottontail	#13
Fox	#15
Black Bear, Jack Rabbit	#16
Coyote	#16
Bobcat, Wolf	#18
Whitetail, Mountain Lion	#23-25
Mule Deer, Antelope, Moose	#26-29
Elk, Buffalo	#31-36

A few representative common birds and their approximate glass eye sizes and colors are given here:

Bird	*Color*	*m/m*
Humming Bird	Brown	2
Sparrow	Brown	4
Starling	Brown	5
Robin	Brown	6
Woodcock	Brown	7
Pigeon	Orange	8
Ringneck Pheasant	Special (3-color)	11
Goose	Brown	12
Cooper's Hawk	Straw	12
Goshawk	Yellow	14
Golden Eagle	Brown	18
Ruffed Grouse	Hazel	9
Mallard Duck	Brown	11
Crow	Brown	11
Great Horned Owl	Yellow	20
Snowy Owl	Straw	22

Body Forms

Prefabricated bird and animal forms made of compressed cork, laminated paper and other compositions are available from various taxidermist supply houses. These eliminate much work and assure excellent results. Consult the various supply catalogs which anyone interested in taxidermy should add to his hobby library. Many purchased forms are easily duplicated by the paper sculpturing methods explained in Chapter 2.

TAXIDERMIST SUPPLY SOURCES AND TANNING SERVICES

Names and addresses of supply dealers will be found in the classified sections of any of the popular sporting magazines. Most of them will be glad to offer suggestions pertaining to taxidermy needs and they will be of great help to the inquirer. Names and addresses of dealers and manufacturers of every description may be obtained from MacRae's Bluebook or the Thomas Register of American Manufacturers, available at any public library, where they are listed alphabetically by subject.

The following list, while not necessarily all-inclusive, offers a starting point for anyone wishing to assemble information on commercial supplies and services of interest to the taxidermist.

Firm	Product or Service
American Balsa-Wood Company Newark, New Jersey 07100	Balsa Wood
American Excelsior Corporation Chicago, Illinois 60600	Excelsior
American Wildlife Studios, Inc. P.O. Box 16030, University Station Baton Rouge, Louisiana 70803	Taxidermist Supplies
Arco Taxidermy Supplies Box 693 Tarpon Springs, Florida 33589	Taxidermist Supplies
Bucks County Fur Products Inc. P.O. Box 204 Quakertown, Pennsylvania 18951	Taxidermy Tanning Services
The Celluclay Company, Inc. P.O. Box 1296 Marshall, Texas 75670	Instant Papier-Mache

Central Scientific Company
Chicago, Illinois 60600

Misc. Specialties

Clearfield Taxidermy
Div. of Clearfield Furs, Inc.
603-605 Hannah St.
Clearfield, Pennsylvania 16830

Taxidermy Tanning
Services; Taxidermist
Supplies; Fur & Leather
Manufacturers

Colorado Tanning Co.
1787-93 S. Broadway
Denver, Colorado 80210

Taxidermy Tanning
Services

Deerskin Leather Co.
420 Chestnut St.
Coplay, Pennsylvania 18037

Deerskin Products and
Accessories

J. W. Elwood Supply Co., Inc.
1202 Harney Street
Omaha, Nebraska 68102

Taxidermist, Tanner and
Furrier Supplies

Jules Garfall, Inc.
10 Glenwood Ave.
Johnstown, New York 12095

Deerskin Tanning and
Products

David Hamberger Inc.
410 Hicks St.
Brooklyn, New York 11201

"Celastic" Plastics

E. L. Heacock Co.
117 Bleeker St.
Gloversville, New York 12078

Buckskin Tanning and
Products

M. J. Hofmann Co.
963 Broadway
Brooklyn, New York 11221

Taxidermist Supplies

Jonas Bros. Inc.
1037 Broadway
Denver, Colorado 80203

Taxidermist and Tanning
Supplies

Jonas Bros. Studios
North High Street
Mt. Vernon, New York 10550

Taxidermy Tanning
Services

Keith Kline
R.D. #2
New Tripoli, Pennsylvania 18066

Tanning Services

Norman K. Meyer 4788 North Bend Road Cincinnati, Ohio 45211	Fish-Mounting Supplies
Penn Taxidermy Supply Co. P.O. Box 156 Hazleton, Pennsylvania 18201	Taxidermist and Tanning Supplies
Rochester Fur Dressing Co., Inc. 219 Smith Street Rochester, New York 14608	Taxidermy Tanning Services
G. Schoepfer 120 West 31 Street New York, New York 10001	Glass Eyes
Smallwood's 28 Waugh Drive Houston, Texas 77007	Taxidermist Supplies
Ernest J. Stalek Woodworking 902 E. Paoli St. Allentown, Pennsylvania 18103	Taxidermist Panels
Van Dyke's Woonsocket South Dakota 57385	Taxidermist and Tanning Supplies

CHAPTER 2

Techniques For Molds, Forms, And Papier-Mache

THE BASICS ON HOW TO make plaster of paris molds, and forms from these molds, are detailed in this chapter. Once familiar with these procedures, one may undertake the fish-casting mounting method, or the making of needed forms in the same general ways, still not neglecting, however, to give special attention to whatever departures from or additions to these procedures are suggested elsewhere in this book for specific projects.

Knowing how to make molds with plaster of paris, and how to make from these molds modeled papier-mache or laminated-paper forms are skills highly useful to the taxidermist. Most taxidermy projects, one realizes, involve mounting a skin over some bird, animal or fish form. While often the form can be obtained from commercial sources, sooner or later the taxidermist will want to make a needed form himself. He may find, in fact, that for some particular job he really has no choice, he will have to make it if the project is to be completed.

If it is to be the kind of paper-laminated form most generally preferred, the first step in its production will require a plaster of paris mold. One also may have to make small accessories now and then, such as ear liners, etc., in which the preparation of a mold is a preliminary step.

While relatively uncomplicated work, some proficiency in mixing and using papier-mache is also basic to most taxidermy work. Not only is papier-mache used when setting glass eyes, but there are numerous other instances when a touch of papier-mache here and there will fill out unnatural form lines and spell the difference between a finished mount that looks right and one that doesn't.

Paper-laminated fish forms also require mold-making skill. Through necessity the taxidermist therefore becomes familiar with using various combinations of papier-mache, papers and glues. In fish mounting work, the ability to make good molds, and fish forms from the molds, allows the taxidermist some choice as to subsequent mounting procedure. He may, for instance, enclose the artificial form in a fish skin to complete the mount, or he may confine the project to one of making a fish reproduction of paper which, when properly oil-painted, will serve as an entirely satisfactory display.

Wholly apart from this, skill acquired in working with plaster of paris and paper sculpture not only adds new dimensions to home taxidermy projects, but it also comes in handy in other hobby work. In model railroading, for instance, skill in working with papier-mache is advantageous, for this substance is well adapted to simulating hills and other terrain. Many other hobbies employ papier-mache.

MATERIALS FOR SHAPING

Assemble the following materials, most of which can be obtained from any building supply dealer.

Molding Plaster

This is sold in large bags and is very inexpensive. The material is mostly used by contractors in molding ornamental architecture in churches, public buildings, etc.

Common Building Sand

Get enough to nearly fill a sturdy wooden box, one large enough to accommodate the largest object to be cast or the largest fish-cast anticipated. The sand should be kept uniformly damp during use.

Paper

Any soft, water-absorbent paper can be used. Common "red rozin" building paper, used by roofers and contractors as underlayment, is ideal. The 20-lb. weight, in rolls, is stocked by most dealers. Use lighter weight for small fish and the 30-lb weight for large fish and heavy form work. Paper toweling, etc., may also be just the thing for modeling fins and producing small details.

Alum and Dextrin

A local druggist can supply powdered alum and several pounds of dextrin, the latter a gummy substance in powdered form obtained from starch and used in preparing paste.

Miscellaneous Items

Other items and equipment needed include white and orange shellac together with a quantity of denatured alcohol for use as thinner and in brush cleaning; an assortment of paint brushes; a two-quart saucepan for making paste; an assortment of mixing bowls, containers, etc.; cheesecloth; putty knife or trowel; papier-mache; plasticene (modeling clay); petroleum jelly; sandpaper; and several varieties of adhesives such as contact cement and common white glue.

Many items such as papier-mache, plasticene and other formerly strictly mail-order items useful to taxidermists are now carried in stock in the art supply departments of large paint and department chain stores.

One must also have the use of a cooking surface for preparing the dextrin paste. A plug-in type electric hot plate is just the thing since it can be placed on the work table and used to keep the paste heated during paper-laminating work.

MAKING PAPIER-MACHE

One can make serviceable papier-mache from soft paper of almost any sort. Shredded newspaper can be ground to a pulp in hot or boiling water by churning and beating it. When the fibers are well separated, smooth the mixture and free it of lumps, squeeze out the excess water and add dextrin paste or glue. To provide body and regulate the setting characteristics of your mache mix, experiment with additions of plaster, borax or potter's clay.

In using papier-mache, there are several ways of strengthening it, one being to spread it over strips or pieces of screening or other supporting material. Always remember to allow plenty of drying time when finishing papier-mache work.

Making papier-mache is not a greatly complicated process, but the task of shredding newsprint can be quite time-consuming. Consequently there is merit in purchasing the commercial product.

MAKING A PLASTER OF PARIS MOLD

A strong wooden box is needed, one large enough to accommodate the work anticipated. Place it on a bench or some other

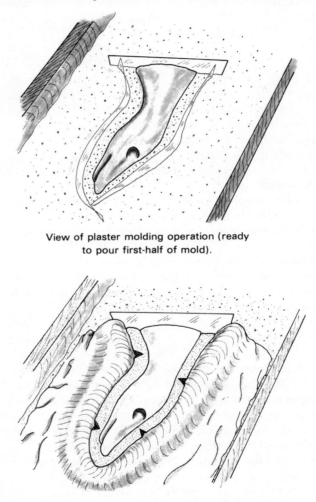

View of plaster molding operation (ready to pour first-half of mold).

View of plaster molding operation (ready to pour second-half of mold).

stable base so that your work will be at an elevation convenient for you. Fill the box three-quarters full of damp building sand.

Determine the median of the object to be cast, in this case an imaginary line dividing the object lengthwise into symmetrical halves, then bury it to this point in the sand. Smooth the sand neatly and accurately around the object.

To keep the plaster from spreading too widely, form a barrier around the perimeter of the object, of a size commensurate with the expected thickness of the finished mold. Sand can be dammed up for this or one can use cardboard or metal strips bent to the shape of the object and pushed into the sand. The exposed part of the object to be cast will form, of course, the first half of a hollow plaster of paris mold.

Apply a separator coating of cold cream, petroleum jelly or similar oily substance to the exposed surface of the object being cast. A thin, watery mixture of potter's clay called "slip" makes a good separator. The use of a separator is necessary to prevent the mold from sticking to the subject being cast.

Estimate the amount of plaster mix needed by figuring two and one-half parts of plaster to one of water. Use a plastic bucket for the liquid plaster since the plastic is more easily cleaned. It will be important that all tools and containers be scoured free of plaster residue promptly after their use. The presence of any old plaster residue affects the setting time of subsequent plaster mixes.

Normal setting time of molding plaster is about half an hour, but this varies somewhat because of factors such as its age, atmospheric conditions, water content, the use of special additives, etc., to mention but a few. One may learn the approximate setting time of plaster by testing it before each new job. From observation and experience one learns to gauge the time of each step accurately.

Having estimated the amount of mix to be produced, fill the plastic bucket with the appropriate amount of clean, cool water. Time now will be of the essence.

Prepare the plaster mixture by sprinkling *fresh* molding plaster slowly over the water. The use of a flour sifter is recommended for this since it helps spread the powder evenly. As the powder is sifted on, allow each layer of powder to be absorbed in the water before adding more. Through slowly and gradually adding plaster, the solution will become saturated until finally the dry powder will begin to stay on top of the mixture. The solution may be stirred gently, or one may insert his hand and work the mixture about slowly with the fingers, adding a little more plaster, if necessary, until its texture achieves the consistency of heavy cream. This procedure helps eliminate air bubbles and produces a good mix. When the plaster texture forms a creamy coating it is nearly ready to use.

Bounce the container, at this stage, against the floor several times to promote the escape of air bubbles. Then allow the mixture to stand and thicken to the pouring consistency desired.

Some mold-makers prefer to brush a thin layer of liquid plaster over the object being cast before pouring. For this, a soft brush of some sort is convenient.

When the plaster mix has thickened for a few minutes it should be ready to pour. Proceed by pouring some of the mix in a controlled flow, applied as evenly as possible over the object being cast. If you are concerned about the possibility of breakage, remember that the mold can be reinforced through adding strips of wood, wire mesh, or some fibrous material such as burlap, excelsior, etc., applied between layers of plaster.

Continue forming the mold by adding plaster mix, pouring on additional layers at short intervals. Though the plaster thickens in the bucket, it can still be applied by the use of a trowel putty knife, or small paddle. Between each layer poured, if necessary, allow a time interval sufficient so that the preceding layer will firm up before the next is applied.

Keep in mind that the poured plaster should form a coat of even, uniform thickness. When the first half-mold has been poured, flatten its top so that it will later rest on a table without rocking.

While waiting for this first half-mold to set, thoroughly scour the plaster remnants from the tools and mixing container. It will prove much more difficult to remove this residue later if you do not do it now.

A freshly-poured plaster of paris mold will emit some heat at first, but this will gradually diminish. When the mold is cold to touch, the whole assembly is dug from the sand, turned over, and prepared for making the second half of the mold.

A little clean-up work is now in order. Remove any sand particles still clinging to the mold or the object being. cast, smooth up the mold's margins by scraping and paring with a knife, then cut V-shaped notches at intervals along the mold's contact edges. This keying procedure provides for accurately joining the mold halves later. Now proceed to grease or soap the object's uncast side, being sure to apply separator also to the mold margin, to prevent its sticking. All surfaces that must later be separated require this treatment.

Rearrange the sand in the molding box, then sink the first half-mold assembly down into the sand until its margin is flush with the sand surface. Check and rearrange the dam originally used to prevent undue spillage, and then complete the second half of the mold by applying a new batch of plaster of paris mix in the same way as done before. Remember to level the top side of this mold, too, providing thus a flat surface that later will not tip or rock.

When this new plaster of paris mold is thoroughly set and hardened, pry the halves apart carefully, remove the object cast, clean it and the molds, trim the mold edges, and put them aside to dry. The cast object now removed, the mold interiors should show perfectly formed details of the subject. After the molds have thoroughly dried, their interiors should receive several coats of orange shellac; two or three will usually do.

MAKING PAPER FORMS

A form model is constructed by pressing layers of paste-saturated paper strips into plaster molds. Paper laminations are thus built up to required thickness in each half-mold, removed when dry, then fitted together to form the desired complete reproduction or form model.

Select soft, absorbent paper of almost any kind for building up paper forms. Paper towel material is good for small casts having fine details to be reproduced. Common "red rozin" building paper used by contractors and roofers is good and especially useful because of its moisture-absorbing qualities. Use it in 15 to 30-pound weights, according to the size of the end product.

Never cut paper intended for use in paper sculpture; always *tear* it. When tearing up such paper, notice that it has a grain running lengthwise, thus it is best torn crossgrain from the roll over a straightedge. For making forms we need narrow paper strips in lengths just long enough to fit precisely to the edges of the mold half after being soaked with hot paste and placed down crosswise in the mold. If, for example, you wish to duplicate the form of a 20-inch fish, hold the strips to a width of about one and one-half inches. For smaller forms keep the strips even narrower. While wider strips are satisfactory for the larger forms, strips too wide tend to buckle along the edges and are more difficult to apply to curved portions of the mold.

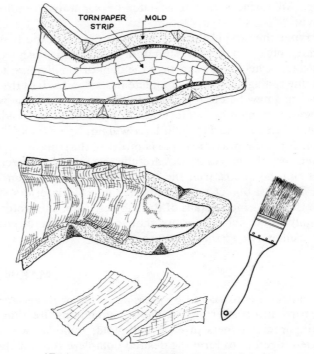

TORN PAPER STRIP   MOLD

(*Top*): Paper laminations completed and drying
in the mold.
(*Below*): The first layer of paper going into the mold.

Making paper forms is somewhat messy work because it involves spreading paste on both sides of the paper strips with a brush. Consequently, the most desirable working area would have a hard, smooth table or work surface available. Enameled metal table tops or formica-covered counters are good since they are less susceptible to stickiness and are more easily cleaned.

The pastes used in paper sculpture are many and varied. The requirements are that such pastes be inexpensive, thin enough to be readily absorbed by the paper, yet have good sticking quality and enough body to dry to a tough, hard substance.

Dextrin meets these qualifications, especially if thickened with a little wheat flour and glue. Some paper workers add various combinations of wheat paste, glue, starch and poster paste (used by advertising people for billboard sticking), and there are other additives, such as gum arabic. Most taxidermist supply dealers stock convenient, ready-mixed pastes in dry form, together with instructions for their preparation. Thin pastes saturate paper best, but thickened pastes are easier to use and dry quicker. It is usually a good idea to employ a fairly thick mixture and apply it while hot.

If the paper form will require any wood anchor blocks, plan the location of these blocks so this will not offer an installation problem later, after the form has already been made. Provide for this, if necessary, in making up the form.

A very thin application of oily separator must be brushed over the inside surfaces of the shellacked mold to allow easy removal of the model after it has dried sufficiently. A coating of thinned petroleum jelly is applied for this purpose.

Assuming paste ingredients in a dry state have been selected for use, mix the ingredients together now, then sift them slowly into boiling water while stirring the mixture vigorously.

The paste now prepared and the paper strips ready, with a wide paint brush apply hot paste to both sides of some paper strips, then press them in so as to form overlapping layers in the mold. Start at one end, pressing the strips into the mold crosswise and with this, the first layer, allow the edges of the strips to extend somewhat above the mold edge along both sides, preferably to a distance of about one inch, or so. This will form a smooth binder-edge when the strips are later folded in. The edges of subsequent layers are kept flush, however, with the margins of the mold. Tamping the softened paper into tight contact with depressed areas of the mold will bring out sharply defined details on the finished form model.

Each layer of paper strips should run in an opposite direction to its predecessor, first a vertical layer, then a horizontal layer, then a vertical layer, and so on. It is good practice to put a thin layer of soft papier-mache in between the second and third layers, especially toward the head end of the form. The mache helps hold the paper in place, provides strength, and prevents curling and loss of detail during the drying phase.

As to the number and thickness of the laminations required, one must exercise his own judgment. Give due consideration, however, to installing any desired wood blocks as anchors for wood screws.

If constructing forms having nostrils, lips, eyes, etc., a layer of papier-mache spread between the paper layers will add thickness and strength. This is advantageous should the finished form involve nailing or pinning, and the papier-mache will help achieve the tight contact with critical details in the mold.

When the laminations have been built up to the required thickness, neatly fold in and press the extended binding edge of the first layer atop the others so that it is exactly flush with the inner edges of the mold all around. It is important that no part of this edge be allowed to extend over the rim!

Allow time for the form halves to dry. This likely will take a day or so, or at least an overnight interval. After the forms have dried enough to retain their shapes, they may be pulled out and allowed to dry some more, flat-side down, slowly and completely.

Assuming that the form will be used in some conventional taxidermy mounting, there are several things to attend to before joining the two halves of the form together. Eye and mouth openings of fish models are now cut out with a sharp blade and nippers. Apply a covering of strong cloth over the inside of the eye openings with contact cement. This forms a backing to later accommodate a bed of papier-mache when the stage is reached for inserting glass eyes. Also, if needed, a small wooden block should be cemented inside the back half of the fish models to provide an anchor for woodscrews, for later fastening the mount to a wall panel. If a pedestal type fish model is desired, that is, one finished on both sides to be displayed on a table instead of a conventional wall mount, now is the time to insert a pair of dowel-support rods. Fasten these to the inside top edge of one of the model halves with a small pin and secure them by wrapping glue-saturated cloth to the dowels and the inside of the paper shell.

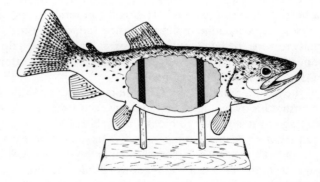

A laminated paper fish model, made up as a "pedestal" mount (both sides for display).

If the form is to be a fish model, start joining the halves by gluing tail, dorsal and anal fins together. Be sure these are brought to proper thickness by additional tapering layers of paper or a filling of papier-mache. An ill-fitting seam, resulting from failure to line up the edges of the paper with the mold margin accurately, may be improved by sawing through the joint with a hacksaw blade. Model a filling of mache along the joint, then glue torn strips of moistened building paper in place. Be sure to feather the edges neatly, thus rendering the seams almost invisible. Common white glue thinned with water is excellent for this purpose.

The pectoral and ventral fins of fish, if having been modeled out in paper, are now cemented on and pinned at proper angles to the body. Model around their bases with papier-mache and trim them to size. Smooth out rough spots, if any, the edges of the fins, around the mouth, etc., with finish-grade sandpaper, then give the finished model several coats of white shellac. A simple fish reproduction job can now be completed by installation of glass eyes and an artistic paint job.

Should this form, however, not require attention to all these details, just join the form halves together with strips pasted along the seams; where possible, both inside and out. Finish by giving the now assembled form several waterproofing coats of shellac.

CHAPTER 3

Mounting Fish

SINCE THE GENTLE ART of angling is the oldest and remains still undoubtedly one of the most popular sports in the world, fish mounting is of first interest to many amateur taxidermists. Unfortunately, there is no "best way," no standard procedure for doing this work. One must simply experiment with suggested ideas to find the methods producing the results wanted.

REPRODUCTIONS

Actually, most fish mounts seen in museums and collections are reproductions. This means that they are cast from molds of the fish bodies and are made entirely of synthetic materials. Even though these are far superior, being free of the shrinkage and deterioration troubles so often encountered in skin mounts, no amount of persuasion can change the mind of the angler who wants the "real thing!" Nevertheless, since fish are particularly suitable for paper reproductions from simple two-piece molds as described in the preceding chapter, it is urged that the amateur fish mounter try at least one of these reproductions. A distinct advantage of this process lies in the fact that permanent fish molds may be used over and over again to produce any number of similar reproductions. Also, the chore of skinning the fish is eliminated!

PRELIMINARY TROPHY CARE

As is true of all aquatic life, fish spoil quickly when removed from their watery element, therefore fish specimens require prompt attention after being caught. It is a good idea to take a close-up color photograph at once, if possible.

Keep fish wet by wrapping them in moist cloths, then in a plastic bag until refrigerated. Fish can be quick-frozen and kept in good condition for a considerable time. In such cases, they must be skinned immediately after thawing, however.

Bass, trout and pike of the fresh-water variety, and some salt-water game fish comprise the majority of fish most commonly considered for mounting. The hard-scaled varieties are the most adaptable for this purpose. Fish of the soft, fleshy type are generally unsuitable for skin mounting and are best reproduced by other means. Both largemouth and smallmouth bass lend themselves admirably for mounting. The pickerel, Great Northern pike, muskellunge, walleye, and similar pikes as well as panfish make good subjects.

Trout require more careful handling, especially if of small size. Large trout make attractive mounts when handled right.

Since there is no standardization in hand-crafted art work such as taxidermy, the methods and procedures in fish mounting are as varied and diverse as the individuals who conceive them. Unfortunately, many persons, having developed new and eventually successful processes in taxidermy, keep the information to themselves.

Shrinkage

All animal tissue is composed largely of water. Thus, skin mounts of fish are subject to this unavoidable circumstance. Compensate for this by applying non-shrinking fill material such as papier-mache, whenever tissue is removed in the head and shoulder areas.

If shrunken areas develop during drying periods, they must be filled in smoothly with papier-mache or lacquer glazing putty on the outside. *Do this only after the possibility of further shrinkage has passed.*

SUPPLY NOTES FOR THE FISH TAXIDERMIST

Here are some special notes for the fish taxidermist relative to preservatives, adhesives, solvents, and containers.

Preservatives

Fish skins can be mounted quite successfully without the use of any preservative. But, of course, it is more practical to protect them with suggested preserving agents as follows: Borax—use as described. Alcohol—pure grain alcohol is best, but too expensive, especially since denatured alcohol (sold in paint stores under various trade names as "Paco", "Solox", "El Kol" etc., for use as shellac thinner, stove fuel, etc.) will serve as well. Mix about two parts of alcohol to one of water for soaking. Alcohol gradually replaces the water in fish skins and hardens the tissue. Prolonged soaking in alcohol tends to shrink fish skins and this may cause trouble in spreading fins and tail so don't overdo this. Formaldehyde (formalin)—Commercial strength is 38 to 40%. It is useful for preserving whole specimens in glass containers. Use 1 part of formalin to 9 parts of water for this purpose. This 1 to 9 solution can also be injected into fleshy parts of large fish, tongue, jaws, etc., with a hypodermic syringe. Use a proportion of 1 part formalin to 15 parts of water for brushed-on applications on the inside of fleshy parts of head and gills and fin junctions. A very small amount (10 to 12 drops) may be added per quart of borax-water soaking solution. Formalin is rather unpleasant to use because of its foul odor and disagreeable tanning effect on the skin of the fingers.

Adhesives

The fish taxidermist has need for all kinds of pastes, cements and glues. Prepare dextrin paste as described in the instructions for paper-modeling work. Dextrin paste is also excellent for gluing skins to wood and paper manikins. Common white glue, a synthetic resin, thins nicely with water, making it easy to manage for paper sticking, etc. It dries clear and forms a tough plastic film when thinly applied. Contact cement may be had in both handy tubes and small bottles with applicators, and has the advantage of quick setting and requires no pressure clamping. One of the best all-around adhesives to use around the shop is Dupont "Duco" household cement.

Solvents and Containers

When one works with such a variety of cements, paints, varnishes, etc., one will most likely require a number of chemical solvents. On the active taxidermist's bench one will find: Spirits, alcohol, acetone, lacquer thinner, carbon tetrachloride, glycerine, paint remover, petroleum jelly, formaldehyde, carbolic acid, lacquer, shellac, turpentine, arsenite, tanning oils, alum, sal soda, and detergent. A great many containers, tins, bowls, small jars (the kind in which baby foods are packed are great) will be needed. One also needs a variety of pins of various lengths. Quarter-inch long dressmaker's "sequin" pins serve many purposes. Also, rust-proofed bank pins, of nickled steel are valuable and may be had in many handy sizes.

TAMPED SKIN MOUNTING METHOD

This method is suggested as the easiest for the inexperienced amateur and requires a minimum of special equipment and materials. Good results are possible even at the first attempt, if the steps are carefully followed.

Preliminary Steps

First, make an accurate tracing of the subject by holding a pencil vertically and scribing its outline on a clean sheet of paper. Note all pertinent measurements and details on the sketch, such as length, girth, and width at several locations, location of eye, lateral line, etc. This information will prove of value after the fish is skinned.

Wash the fish in running water, then wipe most of the slime away with paper towels. A little powdered alum dissolved in water does a good job of removing the remaining mucus. Apply it with a somewhat stiff-bristled brush, not forgetting the inside of mouth and gill structures.

Skinning and Fleshing

Sprinkle water on the top of the work table and lay the fish "show-side" down. Make an incision with the knife and shears along the center line (not the lateral line) of the fish from its gills to tail. Make another incision, vertically, just ahead of the junction of the caudal fin (tail) as shown in Figure 1.

The forward end of the incision should terminate at the hinge found in the shoulder bone, just under the gill covers. Disjoint

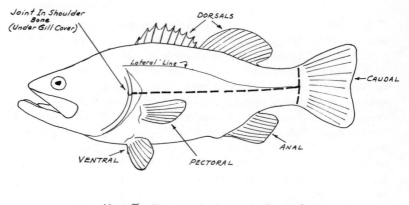

MAKE TWO INCISIONS AS SHOWN BY DASHED LINES

Figure 1.

this hinge. The incisions should be made just deep enough to separate the skin. Avoid cutting into the under flesh as much as possible. Now loosen the skin from the flesh along the upper edge of the incision and peel it towards the top until the dorsal fin is reached. Work the skin free of the body with the fingers, assisting with occasional strokes of a dull skinning tool where necessary. Do the same with the lower half, separating skin from the body along its entire length until progress is halted by the pectoral, ventral and anal fins on the bottom. (See Figure 1.)

With the shears or nippers, cut the bones extending downward from the dorsal fin, just under the skin. A good way to do this without accidentally cutting through the skin on the opposite side is to press the finger tips of the left hand against the skin from the back so that the movements of the instrument can be felt while cutting.

When skinning downward from the incision, try to avoid tearing the abdomen. It will be found that the skinning operation is more difficult around the soft under parts of the fish. Cut the bony structure of the bottom fins just under the skin and also the anal canal. *At the forward end, allow the skin to remain attached to the shoulder bones.* These bones are attached to the gills. Cut this attachment under the gill covers and lay them back as shown in the sketch.

The next step is to sever the body from the caudal or tail fin with the nippers. Use the same method of forcing the point of the instrument between the skin and flesh and guiding with the fingers from the opposite side. Cut as closely as possible to the base of the tail.

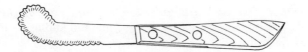

Fish-skinning tool.

Now, since all of the bony connections of fins and tail have been severed from the body, it is an easy matter to skin the opposite or "show side." It is best to leave the fish in the same position ("show side" down), and work the body out of the skin. Work from the tail end forward, holding the skin down with one hand while working the body away from the skin with the other. (See Figure 2.) Avoid stretching the skin at all times!

Figure 2. Fish half-skinned—incision side.

The advantage of this procedure is that the skin is not folded or bent to any extent, thereby avoiding the loss of scales, these being dislodged only if the skin is bent or twisted by careless handling.

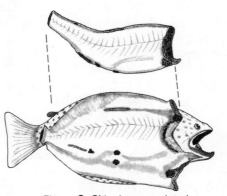

Figure 3. Skinning completed.

Assuming now that the body has been entirely freed of the skin up to its attachment at the head and throat, sever it from the head where the spine enters the body structure containing the brain. *Do not cut the narrow connection of the skin to the gill structure at the throat, and leave the gills, tongue, and part of the gullet intact.* The body meat may now be cleaned and laid aside for food purposes, if desired.

The next step is the removal of all flesh from the head and any that may remain on the skin. Cut out the lower portion of the skull with the nippers and remove the brain tissue with a small scoop fashioned from a piece of wire. Remove the eyes by cutting around their edges and pulling them out with forceps. Remove the soft flesh around the brain case and back parts of the head. Do not cut out too much of the cartilaginous material that forms the skull, jaws and roof of the mouth as this causes excessive shrinkage during drying. Remember that much of the material forming the fish's head is cartilaginous in nature and will dry to a solid substance.

Use a variety of scrapers in different shapes and forms to facilitate this tedious work. A quantity of soft tissue will be found directly under the eye sockets of most fish. Remove this cheek meat by inserting the point of the knife and pressing downward from the inside of the eye socket. Then scoop out as much of the flesh as possible by the combined use of the fingers exerting pressure on the cheek from the outside while scooping out meat thru the eye opening.

When satisfied that all or most of the meat has been removed from the head, turn your attention to the tissue remaining on the skin proper. Scrape the flesh from the shoulder bones and lower part of the throat skin. Be careful not to break the connection between throat skin and lower jaw, and leave the gills and tongue intact. Extreme care must be exercised at this critical point as the throat skin is thin and delicate. Try to avoid removing the white linings of the skin in this area. Cut protruding bones close to the skin, and be sure to remove all flesh from the shoulder bones and bases of the fins. A pointed scissors is helpful in this work. Scrape the entire skin forward, from tail towards head.

When this exacting fleshing chore has been accomplished, the skin should be soaked several hours in borax water. Prepare this brine by sprinkling enough powdered borax in water to make a saturate solution; that is, add enough so that some remains undissolved on the bottom of the container after stirring. If the scales show a tendency to looseness, or if the skin is in poor condition for some reason, add a little salt or alum to the borax-water solution, or, soak for a time in a 65% solution of alcohol.

Assuming that the fish has been properly skinned, fleshed, and soaked for a time in borax water, proceed as follows:

Remove the skin from the wash and wipe the inside fairly dry. Then rub dry borax powder over the inside of the entire skin and head. Use plenty of powder, taking particular care to cover all parts of head, gills, cheek pockets, etc.

Sewing and Sand Packing to Shape

Now, with needle and well-waxed thread, sew up the incision, starting at the tail end. Use a rather close stitch, drawing each up tightly. The best stitch to use is that in which the point of the needle is always inserted from the under, or flesh side; i.e., from the inside, out. When the incision has evenly been sewed up to a point near the center, knot and cut the thread. Now start at the head end and work back. Here it is well to start by punching a small hole through both overlapping ends of the shoulder hinge and then tying them together with the thread.

Now sew the incision back towards the tail to a point one or two inches short of the place where the first seam terminates. This will provide a gap or opening for filling purposes.

Tear a long strip of cloth, several inches in width, and fold it up accordian-wise. Pack this cloth thru the mouth into the shoulders to form a plug at this point, adjusting the cloth to cover all openings. Next, pack a quantity of excelsior into the cheek pockets where the flesh was removed, then fill out the eye sockets with a small ball of the same material.

Now proceed to fill the skin with sand thru the opening left in the incision. Be sure the sand is clean and perfectly dry throughout. Use an ordinary tin can with its rim bent into a narrow "V" to facilitate pouring the sand thru the opening. When the skin has been completely filled and will take no more sand, sew the rest of the opening shut.

Lift the sand-filled fish carefully from the work table and place it right side up on a drying rack.* If the incision has been carefully sewed and the shoulder plug inserted properly, little or no sand will leak out; if it does, act quickly to seal the leak!

---

*(See accompanying sketch.) To insure even drying of the fish skin after mounting, some sort of drying rack must be provided for. A metal grill such as found in most cook stove ovens or refrigerators, consisting of metal cross-rods on a rigid frame, similar to that shown on the sketch will do, or a suitable rack can be made up of lumber and sturdy wire mesh. The idea is to provide a flat, non-sagging surface on which to lay the fish, around which the air can circulate freely.

Suggested Drying Rock

Now pose the fish in the position desired, referring to your sketches and measurements.

The body may now be shaped properly by patting and rolling it with the hands. Position the head and tail properly by placing quantities of excelsior underneath as a support. If desired, the gill covers may be flared slightly, exposing the gills. Smooth these out and prop them open by inserting narrow strips of waxed cardboard between each gill segment.

Most game fish look best with jaws open. Curve and arch the mount to avoid the rigid effect apparent in many poorly-done jobs.

At this point sponge the skin lightly with clear water, washing away any sand, borax, or other foreign matter adhering to the skin. Hold all the fins in proper position with pieces of cardboard cut to shape, each slightly larger than the fin. Use two pieces for each fin, one on each side, fastened with a series of ordinary straight pins. (See Figure 4.) It is suggested that the pieces of cardboard

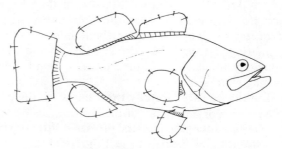

Figure 4. Showing method of "carding" fins.

used for pinning the fins be given a thin coat of ordinary household wax to prevent any sticking when they are removed later.

When all this has been accomplished, let the fish dry undisturbed in a warm, airy place.

When the fish is thoroughly dry, as evidenced by a flinty hardness of all parts, particularly the interior of the mouth, etc., proceed with the mounting as follows: Remove all pins and fin cardboards. Then cut a rectangular section about three inches in

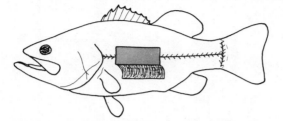

Figure 5. Showing opening made in skin shell.

length and one and one-half inches in width in the center of the back side with three cuts, forming a sort of window through which the sand filling may be removed. Do not remove this section, but fold the skin tab down as shown in Figure 5. This opening may vary in size, depending on the size of the fish. Remove also the cloth plug from the mouth and throat. A lot of material that clings to the interior of the shell may now be removed by inserting a handful of small pebbles and shaking them around the inside. When the inside of the shell has been freed of all loose particles, give the entire skin two coats of thin white shellac, both inside and out, allowing complete drying between coats. Coating the inside may be satisfactorily accomplished by pouring a quantity of the shellac through the opening and tilting and turning the shell until all the inner surfaces are covered.

It is important that the shellacked skin be allowed to dry completely before proceeding further.

Packing With Excelsior

Now fill the shell with chopped excelsior, tamping in small quantities at a time with a piece of dowel or other satisfactory stuffing rod. Fill the skin shell completely, packing tightly, but avoiding excessive pressure. *Beware!, it is very easy to cause depressions in the shell if pressure is put on the outside, so exercise care when handling it because any dents are almost impossible to remove.* When the tamping work is completed, fold the rectangular skin tab back in place and secure with several stitches.

If it is planned to display the mount on a panel, prepare a series of small wooden blocks from three-eighths inch stock which, when

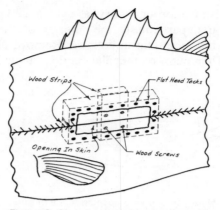

Figure 6. Showing wood insert assembly.

assembled and fastened together inside the fish, will provide a base for fastening the mount to panel, as shown in Figure 6. Make two of these strips about one inch longer than the opening cut into the back of the fish, and wide enough so that when placed side by side, they will permit a one-half inch outer overlapping of fish skin at top and bottom. The third wooden strip will be placed crosswise beneath these two strips and be fastened to them with several small wood screws. To assemble this arrangement within the fish, first lay the crosspiece vertically on the excelsior filling, then slide one of the other strips in lengthwise, centering it and working it upward until the other strip can be slid in below it. Center the two crosspieces on the vertical block beneath and fasten with wood screws.

The edges of the skin should overlap the wooden insert about one-half inch at all points. Staple or tack the skin to the wood.

Completing The Mount

Now turn your attention to the fins. Press a backing of painter's masking tape to all fins and the tail. This is to strengthen and protect them from breakage. Some taxidermists prefer a backing of cloth gauze or cardboard applied with glue. Others strengthen these parts with liquid plastics, etc. Common white glue is good for this purpose. Trim the edges carefully.

Assuming that a glass eye has been properly painted as suggested in the discussion of glass eyes in Chapter 1, it may now be bedded in place with mache. If some difficulty is encountered in seating the eye perfectly a sharp instrument pushed thru from the off side eye socket will aid in proper alignment. In most fish mounting, only the "show-side" of the fish requires an eye.

Complete the interior of the mouth by modeling a quantity of papier-mache or other plastic material where needed. Fish mounted by this method are hung on the wall by inserting a double-pronged wire hanger fashioned as shown in accompanying sketch, if they were not provided with the wooden insert for paneling.

Hanger-wire for hanging fish mounted by
Tamped Skin method if panel is not used.

The mount is now complete except for the restoration of natural colors with brush and oil paints. (See Coloring Mounted Fish, later in this chapter.)

BLOCK MOUNTING METHOD

Panfish such as bluegills, crappies, and calico bass are easily mounted by this simple method. The skins of these small fish, being firm and hard-scaled, dry to a smooth, tough surface making it practical to mount them over carved artificial body forms.

While suggested for panfish, the Block method can be applied with equal success to other species. In fact, the practice of using hand-carved wooden forms in fish mounting is probably the oldest and most common in existence. The chief fault of this method is the likelihood of expansion and contraction of the wood due to changing atmospheric conditions. This is the reason for padding the block with a resilient cotton liner. Also, one is limited to comparatively straight, in-line attitudes and a lot of guess work as to body proportions. The real secret for success with the Block method lies in holding to an accurate *outline* of the fish body. Cut down the thickness as much as necessary to maintain the outline. Also, be sure to adequately waterproof all wooden body forms.

Preliminary Steps

Make the usual accurate sketch of the specimen's shape by laying the fish on a clean sheet of paper and tracing its outline with a pencil held vertically. Note measurements accurately.

Filling Block

Prepare a block similar to the fish's outline, making it just a little smaller and somewhat thinner. Whittle this from a piece of white pine, balsa or similar soft wood. Use the outline sketched earlier. Smooth the carving with sandpaper, then coat it with shellac.

Skin out the body in the usual manner and scrape it clean. Remove the eyes and as much soft tissue from the head and fin bases as possible. Remember to remove the cheek muscles found directly below the eyes, filling in this cavity later with clay or papier-mache. Remember also that much of the tissue forming the head and jaw is cartilage which will dry to a solid substance. Too much removal of this causes excessive shrinkage and results in a poor mount.

Soak the fresh, cleaned scraped skin in a borax-water solution for a short time. If desired, the skin may now be preserved for later mounting by freezing it or by immersing it in a bath of carbon tetrachloride. If the latter is used, wash the skin in clear water after soaking; then salt the skin and allow it to dry. It will keep until ready for mounting, when it can be made pliable again by soaking.

Padding and Finishing

Pad the whittled block evenly with cotton and insert it in the skin. Make sure the core fits tightly, then sew up the incision, finish the head, card the fins and tail, and allow the work to dry. When completely dry, give the mount a coat or two of thin white shellac, allow it to dry, and restore the colors with oil paints.

CASTING MOUNTING METHOD

Fish taxidermy is difficult and time-consuming, in whatever manner it is done, make no mistake about that! For serious work with trophy fish, one must certainly start by making a mold of the fish body as soon as possible. Since shrinkage and deterioration begin the moment a fish dies, the most accurate record is assured by this means. Thus, while more difficult and involved, the Cast-

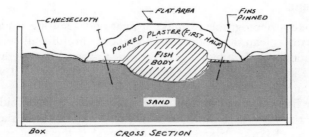

FIRST . . . Apply Plaster To Posed Fish Bedded
In Cloth-Covered Sand.

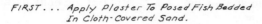

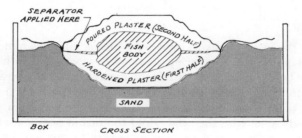

THEN . . . Turn Assembly Over, Apply Separator
To Mold Margins And Pour Second Half.

Making the molds in the Casting method of fish mounting.

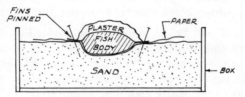

Section diagram—the Casting method.

ing method is indicated for most lifelike skin mounts and is the one probably used by the majority of professionals.

This method results in a most accurate reproduction of the living fish. It requires more time, materials, and experience and gets one into the molding game. The production, however, is well worth the effort.

To mount a fish via the Casting method, we first make a plaster of paris mold of the fish. When the mold is ready for use, it is lined with the fish skin and the skin is then filled with papier-mache. That's the general sketch of what's involved, so we may proceed with the details.

Preliminary Steps

First, take a photograph or make a neat and accurate tracing of the fish on a clean sheet of paper and record carefully all pertinent information as before indicated. By doing this, we become familiar with the subject's contours. Clean the fish by sponging it with a weak alum solution.

If the fish is slack in the gut, make an incision in its off side and remove the viscera, refilling the cavity with any fine material such as powdered borax or sand. (Use cornmeal if the meat is to be eaten.) Fill in enough material to firm out the belly naturally, then sew up the incision.

Making the Mold

Follow the instructions for making a plaster of paris mold as explained in Chapter 2, covering the sand, however, first with a piece of cloth before placing the fish in the desired mounting position on the cloth with the incision side down. The cloth will keep the fish free of sand. Press it down until its tail and top and bottom fins are nearly level with the sand. Hold the fins in position with long straight pins. In some cases it is necessary to remove the "showside" pectoral and ventral fins for the casting operation, pinning them back in place later when the mounting is completed. At this time the fish must be posed accurately in the position desired, because the position cannot be altered once the cast is made.

If the head will be included in the mold, the mouth and gill covers, if spread, will have to be filled with modeling clay to prevent plaster from entering these openings. Fill them completely and smooth the material neatly to the edges of the mouth and gill covers so as to avoid any undercut situations.

While the plaster of paris is made as described earlier in Chapter 2, it is advisable to add to the mix about one-half teaspoon of powdered alum for every quart of water. Addition of the alum tends to harden the mixture and neutralize any mucus remaining on the fish skin. Make sure the body surface is wiped clear of water and mucus before applying the plaster.

Since you will be using papier-mache to help fill out the fish body on the reverse you will need to make only one mold now, pouring the plaster of paris carefully over the fish body from the eye to and including the fins and tail. A rather thick covering is desirable. Complete the mold, trim and smooth it, allow it to dry thoroughly, and treat its interior with a coat or two of orange shellac.

Skinning the Fish

Now skin out the fish as described before, clean it thoroughly and prepare the skin by dipping it in a 65% denatured-alcohol solution. The kinds sold in paint stores as shellac thinners and fuel, under various trade-names can be used by adding two parts of the alcohol to one of water. Saturate the fish skin well in the solution for an hour or so, then drain. Do not allow the skin to remain too long in the alcohol bath. Fish skins should never be soaked for long periods in any liquids. A formaldehyde solution made by adding 15 parts of water to one of commercial formaldehyde, may be used on large fish or on extremely oily and fleshy specimens. Brush the solution carefully only on the fleshy parts, not on the skin, as formalin may have a deleterious effect on fish scales.

Mounting the Skin

When this work is completed, press the skin into the prepared mold, adjusting it to fit perfectly. Mix a quantity of dry-form papier-mache by adding cool water to the powder. Blend this mixture to a rather soft consistency and plaster it carefully in layers inside the skin. Before the mache hardens, it must be pressed solidly in contact with all parts of the mold. Make sure there are no wrinkles in the skin. Set a block of wood in place within the mache body for use as an anchor for accepting wood screws later. Now model the rest of the skin with thickened mache and bring the incision together to a neat fit.

When the mache has hardened to some extent, sew the incision up as described before. Obviously a fish of any appreciable size would be objectionably heavy if entirely filled with mache. To compensate for this, make insert blocks of lightweight balsa wood. Styrofoam® may also be used for this purpose. These blocks can be carved from one or several pieces, to such shape as will fit within the mache-lined skin. Make sure, however, that these inserts are placed well away from the "show-side" of the mount.

It must be remembered that during this procedure the skin must not be disturbed in its contact with the mold, since to do so would affect the shape of the finished mount.

When the mache is completely set and firm to the touch, remove the mount from the mold and fasten it to a temporary panel with wood screws. See accompanying fish supporting gadget sketch, the item made from a common goose-necked desk lamp.

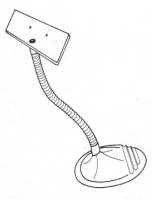

Homemade mounting support especially for fish painting, etc.

Completing the Mount

Complete the mount by carding the fins, adjusting and finishing the head, mouth and gills as described. Allow plenty of time for the mount to dry thoroughly before applying the final oil-painted finish.

Fins and tail, projecting as they do from mounted fish, are vulnerable to accidental damage. Reinforce these by applying a backing cloth, cardboard or fiber with common white glue or "Duco" cement. The light cardboard (oaktag) or heavier red fiber-stock, used in file folders and notebook covers available at stationery stores, is excellent for this purpose. Split or damaged fins are easily repaired by the same means, especially if an additional application of strong thin tissue (eight-pound) is made on the front side. This "onionskin," as it is called, should always be torn to size, providing feathered edges for making invisible repairs and patches on the skin.

Many skin mounts can be strengthened and improved by filling in depressed areas at the fin junctions and head, etc. which result from shrinkage during the drying process. Use papier-mache or, better, lacquer glazing putty (made by American Lacquer & Solvents Co. of Hialeah, Florida) for this purpose. The putty is thinned with lacquer thinner for brushed-on applications. *Do this only after the possibilities of further shrinkage have passed.*

PAPER FORM MOUNTING METHOD

Practically all modern professional taxidermists use paper forms for game heads and the mounting of whole animals. These forms

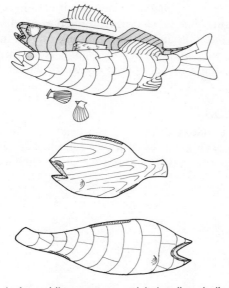

(*Top*): Assembling a paper-modeled walleyed pike.
(*Center*): Carved wooden-block body for a bluegill.
(*Bottom*): Laminated paper form for perch.

are either purchased from dealers or fabricated in the shop by methods quite similar to those described in Chapter 2. A great many taxidermists follow the same procedure for mounting the skins of fish because of their skill and familiarity with the paper modeling process. The resultant mounts are extremely true to life, strong, durable, and light.

Use Molds and Paper Forms

To mount fish by this method, prepare the necessary plaster molds and the paper fish form, following the techniques described in Chapter 2. Note that the head, tail and fins are not required in the form. Also, when making the form provision must be made for the thickness of the fish skin, hence the paper fish form must be made slightly smaller. Do this by gluing layers of paper to a thickness of about 1/16 of an inch or so in the interior of the mold to compensate.

Completing the Mount

When the paper form has dried, gouge out small depressions where the bases of the fins will be seated, if necessary, and treat the form with several coats of shellac. The prepared skin is fitted over this form and papier-mache smoothly applied in head and throat areas where required. Gill and head details are pinned in place, the fins are carded, etc., and the whole is put aside to dry completely before receiving a finishing coat or two of white shellac.

MOUNTING FISH HEADS

It is a simple matter to remove the eyes and clean all the meaty tissue from a fish head, soak it for a time in salt water, dry it completely, then coat it with clear varnish or shellac. However, the

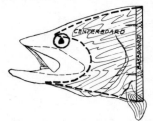

Showing fish-head mounting detail.

enterprising amateur aspires to do a more workmanlike job and devises methods to produce more attractive results.

After measuring up the specimen, he decides to include the shoulders and pectoral fins, and cuts out a wooden base block accurately fitting the girth of the fish. He then fastens a centerboard at right angles to the base block, shaped to fit the throat and notched to seat neatly within the skull cavity, as shown in the sketch.

Cleaning the fish head, but allowing tongue and gills to remain, he soaks the specimen in a preserving solution of 15 parts of water to one of formaldehyde. He then rinses the head in clean water and bathes it in a saturate solution of borax water.

The block assembly is built up to shape with cord-wrapped excelsior or tow covered with mache, and the head fitted in place.

Flared gill covers are pinned in position and the gills spread and held in place with waxed cardboards or flat strips of balsa wood. Balls of excelsior are packed into critical openings to avoid shrinkage while drying. When the head has dried completely to a flinty hardness, glass eyes are bedded in their mache-filled sockets, the interior of the mouth finished with a carved wooden tongue, and the mount completed by restoring the fish colors and fastening it to a decorative wall plaque.

MOUNTING LARGER FISH

Few amateur taxidermists get the opportunity to work on the larger varieties of game fish. However, those who live in coastal areas may wish to try their hand at mounting sailfish, tarpon, or others of the more attractive game species available to them. Each type presents an interesting challenge, the various problems providing endless opportunity for experiment and practical thinking.

The old method of building up an artificial body by using a board shaped to the outline of the fish, wrapping it with cord over quantities of excelsior and tow and covering it with clay, has largely been superseded by more permanent and practical methods for mounting large fish.

Some modern taxidermists are quite proficient in preparing laminated paper forms. These are made by casting both sides of the body in plaster of paris, forming a hollow mold of each side. This can be done on the open beach if the weather is good. Layers of dampened paper coated with adhesive are next pressed into the molds and allowed to dry. The halves are then removed from the mold and fitted together to make an exact duplicate of the fish form. The skin is then removed from the carcass, preserved, fitted, pasted to the form, and finished in the conventional manner. This manikin method can be used on any sized fish and results in a desirable lightweight mount. If used, however, it must be remembered that the plaster molds must be lined so as to compensate for the thickness of the specimen's skin.

The following method should be somewhat easier for the amateur, yet it is similar to the Casting method described before.

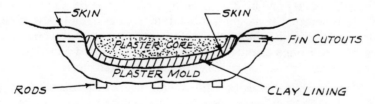

Section view and details of large-fish mold.

The pectoral, ventral and dorsal fins are cut from the body and kept in a borax or alcohol solution until later. The fish is then posed on a large flat piece of plywood (or in cloth-covered sand as before) in the desired position for mounting. A heavy coating of rather thick plaster of paris molding mixture is applied with the hands over the fish from the middle of its head to the junction of the tail fin. After the plaster has set, the cast is removed and the fish skinned.

The process of skinning a large fish is essentially the same as described for smaller fish. The gills are removed and the skin is scraped clean. All flesh is removed from the head, fin joints and tail junction. If the skin and head are greasy, they are bathed in naphtha or other solvent, then rinsed in borax water. This is a "must" if the fish is to be mounted properly. The inside of the drained skin is then rubbed down with salt until fairly dry. This facilitates fleshing and curing of large skins.

The mold is reinforced with wire mesh or wooden strips to avoid breakage. Openings are cut out along the edges of the mold to allow the fins to be set in place.

The skin is now cleansed of salt and preserved with a borax water solution. Plenty of dry borax is rubbed into the head and other critical parts. The skin is next accurately fitted into the mold and secured in tight contact with it by pressing on an overlayer of water-softened potter's clay. (This material may be purchased in dry form from any taxidermy or artists' supply house.) This clay lining is made only about one inch thick, depending more or less on the size of the specimen.

A batch of plaster of paris is next poured into the rest of the opening to form a solid core which holds the clay in place until dry. After this complete assembly has dried, a wooden panel is placed on the back and the assembly is turned over and the mold removed. The tail and all fins still attached to the skin are positioned properly and held in place by carding and pinning them to the baseboard. The mount is now left to dry out completely. When dried to a flinty hardness, the mold is replaced over the body and turned over once again. The plaster core and dried clay liner are easily removed and the inside of the skin brushed clean.

The dry skin shell is now filled with layers of prepared papier-mache, each layer being allowed to dry before adding another. A wooden block is shaped and set in place with more mache, becoming a base for fastening the mount when hung. The skin incision is then closed by sewing, and the whole mounting is again set aside for final drying. Fins which were removed before the fish was skinned, having been carded and dried in proper shape, are backed up with cloth, cardboard, fiber or other material, and installed on the mount with wires or pins and their bases modeled in with mache.

Special attention is given to the dorsal fin of the sailfish. Some taxidermists glue this large fin to a thin sheet of plywood where it is held in place with pins and cardboards until secure. The wood is then trimmed to the contour of the fin and beveled back. The wood is left extended on the bottom, providing a base for fastening to the center block with screws. Most professionals, however, prepare an artificial fin. In fact, many of them lean more and more to artificials and replica models for large fish.

Depressions and openings about the face and head and any others forming during the drying periods are filled in and smoothed with papier-mache.

This method can be varied somewhat by making a hollow body, resulting in a lighter mount. When the dry skin shell has been lined with a thin layer of mache, laminations of pasted building paper are built up on the inside through an elliptical incision cut large enough to admit the hands. These paper laminations are brought to a thickness commensurate with the size of the mount. On large jobs, further reenforcement can be accomplished by applying layers of Celastic and cross-ribbing of the same material.

Celastic, incidentally, is a tough, waterproof plastic, impregnated on fabric, made by the Celastic Mfg. Co. and available thru Ben Walters, National Representative, 4600 E. 11th. Ave., Hialeah, Florida 33013. Swatches of the material are torn to desired size and dipped momentarily in an acetone solvent which makes them immediately self-adhesive, soft and temporarily pliable for molding. In a few minutes the Celastic stiffens to a point where it can be formed free-hand to various shapes, none requiring supports. In about a half-hour or so, the plastic takes on a stone-like hardness that permanently holds its sculptured shape. This material has many qualities that make it useful in taxidermy in such areas as form making, ear liners, fish work, and the like.

COLORING MOUNTED FISH

Restoring color to mounted fish seems to be a major stumbling block for most taxidermists, whether professional or amateur. It goes without saying that this technique must be learned, because no fish retains its natural colors after being taken from the water.

After skinning, even the markings disappear and the fish skin becomes dull and colorless. Such is the case with all aquatic life.

The job of painting is made easier if a good color plate of the species is used as a guide, because rare indeed is the person who can paint a fish from memory! The ideal situation, of course, is to have for reference an actual color photograph of the fish, taken promptly when caught.

Good color prints of game fish which also can be used for reference are to be found on the covers of sporting magazines or in books available at public libraries.

Oil painting is the best method for the home taxidermist to use in coloring mounted fish. However, this work is sometimes done by others using various lacquers and synthetic enamels applied with expensive spray-gun equipment.

Art stores and libraries carry books giving the basic principles of oil and airbrush painting which should be studied by the taxidermist inexperienced along these lines.

The selection of good brushes is important. Purchase the best grade, square-type bristle and pointed-sable brushes in as few sizes as necessary to do the job. Other sizes and types can be obtained later as the need arises. Avoid cheap substitutes, as these lose their shape and lack the flexibility of the better types.

Use the same judgment in acquiring tube oil paints, buying only necessary basic colors and filling in the rest as required. It is a good idea to buy the large studio-size tubes, thus avoiding the necessity for skimping on the colors used most frequently, as is apt to happen if only the small tubes are in hand.

In painting fish, it will be found that a greater quantity of white is used than any other color. For suggested colors, see Chapter 1. Linseed-based drying oil is about the most satisfactory conveyor for the tube-oil pigments. This can be purchased in bottles and kept in a small tin with the palette where it is instantly available.

An annoying characteristic of oil paints is their infernally slow drying time. This can be speeded up considerably by using the quick-drying titanium white, or one of the clear, colorless drying mediums put out under all leading brands. Colors may also be diluted with copal painting medium which accelerates drying time. For the person in a hurry (and who isn't, these days), acrylic paints are the answer. However, these quick-drying paints are more difficult to apply and are not easily stippled.

Get the paper-type palette which is padded in a tablet form and shaped in conventional palette form. These are very handy and

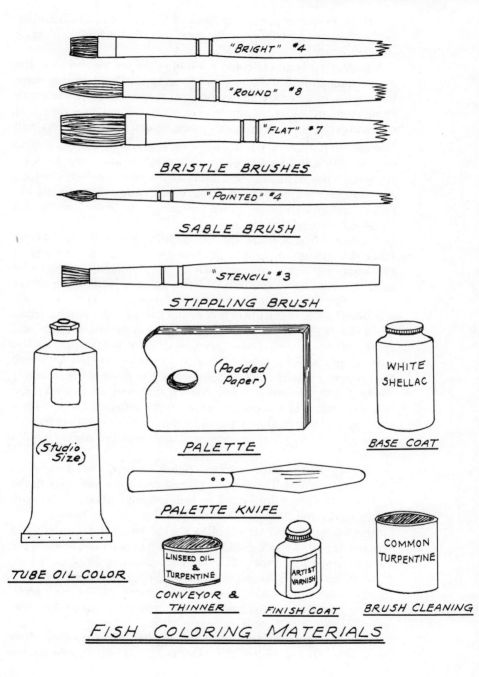

"BRIGHT" #4

"ROUND" #8

"FLAT" #7

BRISTLE BRUSHES

"POINTED" #4

SABLE BRUSH

"STENCIL" #3

STIPPLING BRUSH

(Padded Paper)

PALETTE

WHITE SHELLAC

BASE COAT

(Studio Size)

PALETTE KNIFE

TUBE OIL COLOR

LINSEED OIL & TURPENTINE

CONVEYOR & THINNER

ARTIST VARNISH

FINISH COAT

COMMON TURPENTINE

BRUSH CLEANING

FISH COLORING MATERIALS

practical, allowing each messy sheet to be torn off and discarded, exposing a clean surface for the next job, thus avoiding a nasty cleaning chore.

A painter's palette knife and a roll of paper towels complete the list of equipment. Keep brushes in good condition by rinsing them thoroughly in ordinary spirits or gum turpentine, then washing them in soapy water, a simple procedure taking little time and well worth the effort. Keep paint tubes capped to avoid drying out.

While it is true that not everyone has artistic talent, almost anyone with a little effort can turn out reasonably nice work by following a few basic rules and doing a little practice. Nevertheless, oil-painting mounted fish is not always the easiest thing. If the job fails to measure up to one's own critical eye, however, the job can be done over again!

A suggested procedure for fish painting might be as follows: Place the fish in an upright position facing a good light and at eye level. Squeeze a blob of each of the dominant colors to be used in a row along one edge of the palette, including a larger amount of white which is needed most frequently. Apply a goodly amount of white to the lower half of the fish with the palette knife. Then dip the wide-bristle brush into the conveyor oil cup and spread the paint evenly. Now mix a bit of color to the white on the palette, and apply a band of this to the fish's middle. Continue the process until the fish's upper parts are painted to their darker shade. Clean the brush between each different application, and mix each color in a separate place on the palette. Then, with a swirling, circular motion, lightly blend the colors together so they graduate smoothly from creamy underparts to the darker shades on top.

Now stipple the entire surface, lightly at first, then heavier as conditions warrant. This stippling is just about the best trick in the book for blending colors and it spreads the oils and does away with brush marks and flaws made in the initial paint application.

Brush bristles used for stippling should be of equal length and somewhat stiff. Artist's round stencil brushes are usually used, but various other kinds such as typewriter cleaning brushes, toothbrushes, etc., may also prove adequate. Stippling is done by pressing the brush lightly on a thin smear of pigment, picking up a small amount of the tips of the bristles, then applying paint to the areas to be stippled with a short tapping motion.

When the base colors have been applied satisfactorily, allow the surface to dry until it is tacky. Then mix and stipple on the more dusky markings evident on most fish. After this has dried, the detail markings, spots, vermicular—or worm-shaped configurations, fin and gill-cover markings are painted in with a smaller brush. The distinct lines of fins, jaws and head markings can be accented by lining them with a pointed sable brush and the sharp tips of brush handles.

The fish mount is now laid away in a dry, dust-free place until it is completely dry. This may take considerable time, according to atmospheric conditions.

If a glossy luster is desired, a coat of artist-grade varnish may be applied. A paint job which turns out to be too bright may be toned down by applying a thin wash of clear matte varnish or conveying-oil to which a small amount of dusky pigment has been added, a process known as scumbling.

Artist-grade lacquers and varnishes in glossy and matte (dull) finishes are available in handy spray cans. Either type of finish may be favored, depending on the preference of the artist.

A recent development in painting techniques has been the advent of acrylic polymer emulsion artist's colors. These have the advantage of being very fast drying and they are conveniently thinned and cleaned up with water. They are available in all standard colors and are handled in much the same manner as the conventional oils.

Acrylics may well be the answer to many who desire a quickly applied and completed coloring medium, especially valuable in production work where time is a factor. The fast drying feature of these pigments, however, may work to the disadvantage of the beginner in fish coloring until he has had opportunity to experiment with application techniques. At any rate, investigation into the possibilities of this modern development may be well worth the effort.

The enterprising artist will find a fertile field for experiment in this interesting work, and he should investigate the possibilities of all kinds of artist's aids such as essence of pearl, flake additions and the like.

Damar varnish is the best final protective coating given oil-painted fish. It imparts a wet-looking gloss better than any other finish. Damar used too soon, however, can hold back drying indefinitely. To protect still wet paint use Retouch varnish spray, a diluted form of Damar which will not entirely stop air circulation, but will keep out dust or dirt. Varnish offers protection for the oil colors from moisture and such chemicals as pervade our modern atmosphere.

To soften or gray any color, one may add small amounts of black, brown or gray, or, mix it with its complementary or opposite color. The basic opposites are yellow and violet, orange and blue, red and green. Zinc white makes cleaner, clearer tints mixed with cool colors. Adding white to a color tends to turn it toward violet. To match a particular color, one must first select the closest color to it one has, then try to mix it by adding only one other color, if possible. It sometimes takes even a third color to get the desired result. But keep in mind that with each color added, a mixture becomes duller, a step closer to black. To darken a yellow without losing its brilliance, use brown. Black is inclined to turn it green.

If the cap is stuck on a tube of paint, don't twist it until the tube is broken. Warm it up around the collar by holding it over a lighted match, then it should loosen and come off.

Finally, it is noted that some persons offering advice on fish painting suggest that colors be kept as thin and transparent as possible, hoping thereby to retain some of the coloration and markings of the mounted skin. It is doubtful that this will be effective. In most cases, it is better to strive for complete color values with full coverage.

CHAPTER 4

Mounting Birds

IT IS COMPARATIVELY
easy to skin most birds and mount them in a lifelike manner except
those that are very small. Bird skins are generally tough and durable,
lying loosely against the solid muscular body, making them easy to
remove without damage. Fats and remaining tissue are easily scraped
away, leaving a clean, easily-managed skin with which to work.
Feather tracts of birds form a smooth canopy over an otherwise
ungainly form, hiding minor imperfections made during the mounting process. Careful preening and adjustment of feather tracts
together with proper positioning of wing and leg attachment points
are of major importance in good bird taxidermy.

It must be kept in mind, however, that even though the mechanics
of bird mounting are easily mastered, the final assembly and posing
of specimens in truly characteristic attitudes require knowledge
and study.

ADVANTAGES OF PROCEEDING SLOWLY

It is only natural for beginning amateurs to be overly eager, but
avoid trouble by proceeding slowly and patiently. Impatience and
haste are the biggest causes of failure in almost any new under-

taking. After some experience is gained, trouble spots will be anticipated and the necessary measures taken to overcome or minimize them.

"Experience is the best Teacher" is more than a cliché when it comes to taxidermy. However, knowing beforehand a few of the trouble spots likely to be encountered in bird mounting will greatly help the beginning amateur.

The first difficulty usually occurs in skinning around the posterior or tail end of the bird, and at the small of the back. The soft abdominal parts, complicated with anus, viscera, etc., and thin skin at this point causes this. Annoying blood and body fluid drainage may be a problem while wing joints are being severed from the body. Control this by using ordinary cornmeal or powdered borax and cotton swabs to avoid soiling plumage. Don't use borax for this purpose if the meat is to be eaten.

Trouble will be experienced in inverting the neck skin over the heads of some birds. If necessary, use the alternative back-of-neck incision as described in the text.

Many amateurs find to their dismay that the artificial bird form they so carefully prepared is too large for the skin; a "word to the wise" here, should be sufficient! Positioning the legs and wings and properly locating their points of attachment must be carefully considered before sewing up the incision. Be sure that the skin is positioned over the body form so that the feathers lay neatly in place, as nature intended. Should some stubbornly refuse to fall in place, readjust the skin on the form until the condition is corrected. Finally, please keep in mind that birds as well as mammals can be carriers of disease germs and other parasites, so avoid trouble by refusing to work on unwholesome subjects, selecting only clean, freshly-killed specimens to mount.

FIVE STEPS IN MOUNTING SMALLER BIRDS

The work involved in mounting birds may be divided into consideration of the five chief phases involved. These steps are each reasonably distinctive whether the project is a mount with outstretched wings or merely a conventional folded-wing mount.

Step 1 - Skinning the Bird

Before the skinning operation is started, plug up the mouth, nostrils, and vent (anus) with tufts of cotton; then, laying the bird on its back, its head to the left, proceed as follows:

Opening incision.

Separate the feathers on the breast and make an incision from the point of the breastbone to the vent. Be careful to cut no deeper than necessary to slit the skin. It will be found on most birds that there is a natural channel of bare skin down the centerline of the breast, a boon to the taxidermist. With the fingers, carefully separate the skin from the body with a pushing action, away from the cut. Continue this separating as far as possible at all points, working down the sides and towards the tail.

The first obstacle encountered will be the junction point of the legs. Work the skin away from this point, around the outside of the thighs, working around the legs, until the finger tips meet. Work the knife point through the point where the leg has been encircled and sever the leg at the knee joint. Do the same with the other leg. It should now be an easy matter to continue skinning around to the back of the bird, always loosening the skin with a pushing rather than a pulling motion.

It will be found on most birds that the skin sticks more tightly to the back than any other place; here, therefore, it pays to proceed cautiously. In most cases it is possible to encircle the bird, completely, going around its body with the fingers; if so, it is an easy matter to skin rearward until the tail appendage is reached. With the nippers cut the tail bone and caudal assembly, just forward of the ends of the tail quills.

If it proves too difficult to encircle the body completely with the fingers as described, proceed rearward from the incision at the anus, cutting the anal canal just above the opening and skinning back to

Figure 1. Legs and tail severed.

the point just forward of the tail quills. Cut the tail bone with the cut-off nippers, then with the knife carefully cut thru the remainder of the caudal bones and flesh until the skin is reached on the back side. It will help to bend the tail backward during this operation. The bird at this point should appear as in Figure 1.

After this most critical part of the skinning job has been accomplished, tie a strong cord around the rear portion of the skinned carcass and suspend the bird from above at a convenient working height, as in Figure 2.

It is now an easy matter to pull the inverted skin downward towards the neck and head. The next obstacle to be encountered will be the wings. Bring the skin down, pushing along the wing muscles toward the elbows. Remove the flesh from the upper arm bones, then cut the wing bones at their junction with the body. It is surprising how far out toward their ends the wings can be skinned in most birds. Do not separate the ends of the flight quills from their anchorage on the arm bones. Particular care must be taken at the wing junction points; the knife must be used at points where the skin will not readily free itself. Wipe away any blood that may flow from cuts made on the body during the skinning operation.

Continue to peel the skin forward over the breast, neck, and up to the head. Be careful not to stretch the skin at all times. Little resistance will be encountered until the head is reached. Here the ear entrances will halt progress. With the point of the knife, carefully cut deeply into the ear canal, severing the skin as far in as possible. Too shallow a cut here must be avoided. With the knife assisting in loosening the skin over the skull, continue skinning forward on the head until the eyeball is reached. Cut closely to the skull with the knife. The cut must be made just to the rear of the eyelid. Actually, the inner lining of the lid is cut, and the eyelid is

Figure 2. Suspending bird for skinning.   Figure 3. Bird skinned completely.

freed from the edge of the eyeball which remains in the skull. The head skin is now worked forward to the mandibles.

The most difficult part of bird mounting has now been accomplished and the result will appear as in Figure 3.

On some large-headed birds such as certain ducks, owls, hawks, etc., it will be found impossible to invert the neck skin over the head. In this case, make an incision at the nape of the neck, back a short way on the top side (Figure 4). Sever the neck and complete skinning the head through this incision, which is sewed up later. If the bird's head is crested or the repair would be visible, make this incision to the off-side of the final pose.

Step 2 - Preparing The Skin

The skin is inverted downward over each leg to the ankle joint. Do this by forcing the skin downward with the fingers rather than by pulling. Remove all the flesh from the legs but retain these bones. Do the same with the wings, removing all flesh as far out towards the wing tips as possible. Stretch the wing out, bottom side up, and there make incisions on the underside of large wings as illustrated, removing as much flesh as possible. (See Figure 5.)

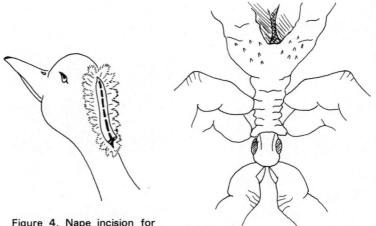

Figure 4. Nape incision for large-headed birds.

Turning head skin back over skull.

With the scraper, carefully shave away all flesh, bits of fat, etc., from the interior of the skin, scraping always forward (towards the head). Remove any flesh and fatty tissue from the tail appendage, being careful not to disturb the roots of the tail quills. In doing this work, never stretch the skin nor pull on the feathers.

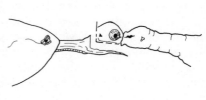

Figure 5. Remove the flesh from the arm bones of large birds by incising on the under side as shown.

Sever head from neck as shown. Then remove contents of head by cutting from the under side of skull.

With the knife, cut an opening in the skull on the underside, exposing the brain — as shown in the accompanying sketch. Remove the brain with one of the wire tools formed into a small scoop from a piece of heavy-gauge wire. Remove the eyeballs carefully, without puncturing them. Clean the skull, removing as much flesh as possible, but do not disengage the jaw bones.

Press papier-mache or modeling clay into eye sockets, skull cavity and throat area. Remember, very little flesh has been removed from the skull, so use the material sparingly. Apply a little soap lather to the skull and neck skin and work the skin back over the head very carefully to avoid tearing. Since there is generally a tight fit here, proceed cautiously.

When the cleaning work has been finished on the inside of the skin, rub dry borax over all parts, inside the body skin, legs, wing openings, and on all retained bones. Rub the borax in thoroughly. If the skin is very greasy as will be found to be the case with most water birds, and others, wash the skins on the inside with pure white gasoline before boraxing them. The feathers should now be thoroughly cleaned and dusted. Complete hand-washing in soap or detergent water is indicated for badly-soiled skins. Have no fear of immersing such skins completely. The wet plumage is easily dried by dusting, first with dry plaster of paris, and then with powdered borax.

Step 3 - Forming the Artificial Body

Place the skinned carcass on its side on a sheet of paper and, holding a pencil vertically, trace around its contours. Then turn the body on its back and repeat, getting an outline as seen from below. These sketches indicate the body size required for the mount. It is also advisable to sketch the shapes of legs, neck, etc., to help in determining the sizes of these parts. Be sure to mark the points of attachment of legs and wings.

Take a ball of excelsior which, when squeezed together in the hands, will approximate the body size required. Then, wind the cotton cord tightly around on all sides, using plenty to bring it up to the required size. Try to duplicate the shape of the natural bird form by squeezing and wrapping until this is achieved. Use the thumbs to put pressure at low points, to taper the tail end, etc. Wind plenty of cord around, making a firm, well-packed body. Remember that the body is generally heavier at its forward end, narrowing to the top and rear. Wads of excelsior may be added here and there to fill out the breast, etc., all firmly wrapped in place with cord. Do not use cotton on or in the body.

When the artificial bird form has been shaped, just a shade smaller than the natural body, wrap it with a piece of wire of proper size, encircling the body lengthwise with it, in a vertical position; clinch this wire at the forward or heavy end and allow a length to extend the distance of the neck and the head plus about one inch. (See Figure 6.)

Figure 6. Artificial bodies formed of cord-wrapped excelsior.

Spin a sufficient amount of cotton batting on the neck wire and wrap it in place with cord to form the neck. This may be made a little thicker than the natural neck. Ascertain the correct length from your sketches. Sharpen the neck wire on the end, since this must be pushed out thru the skull. The artificial body is now complete and should appear as in Figure 6. Body forms for the smaller varieties of birds can also be fashioned from balsa wood. It will be found somewhat easier to duplicate details of the diminutive bodies by such carving. Complete the wooden form by sanding it and coating it with shellac.

Step 4 - Leg and Wing Wiring

See Figure 7. Cut a wire of correct diameter, pointed at one end and twice the length of one leg from the foot to the end of the thigh bone. Force this wire thru the ball of the foot, up thru the back of the lower leg until it appears at the knee joint, inside the skin. Draw it through to a point just beyond the end of the thigh bone. Tie the leg bone and wire loosely together with cord. Then wrap cotton batting around the bone and wire assembly, and bind it in place by wrapping sufficient turns of cord to provide a smooth, evenly-tapered thigh of the proper thickness and diameter. Keep it just a trifle smaller than the original. Repeat this operation with the other leg.

The foregoing work will be simplified by first removing the leg tendons in the lower leg by slitting the bottom of the foot and then pulling the tendons out with a pointed hook. This provides room for the leg wires and makes the job easier.

While it is not absolutely necessary to wire the wings, except where they are to appear spread in the mount, it is recommended. The procedure is much the same as in the leg wiring. Prepare pointed wires, double the length of the spread wing from the tip of the hand bones to their junction point with the body. Insert the wire at the wing tips, draw it through underneath, within the skin opening and out to the end of the upper-arm wing bones. Wing

Figure 7. Leg and wing wiring.　　　　　Figure 8. Final assembly.

bones should be wrapped in the same manner as the legs. Any openings that were made in the undersides of the wings for removing flesh should be well rubbed with borax, padded lightly with cotton, and sewed shut, if the wings will be spread and their inner surfaces be exposed.

Assuming a trial fit has shown the bird form to be of correct size and the breast incision closing together handily, the time for the final assembly is now at hand.

Step 5 - Final Assembly

Insert the pointed neck wire of the artificial body up thru the neck skin to the center of the bottom of the papier-mache-filled skull. Push it thru the skull and out the top of the head. Then push the head down the neck wire until the end of artificial neck is forced tightly against the bottom of the skull. Draw the skin carefully down over the neck and breast.

Next, start one pointed wing wire into the body at the proper position, its natural junction, and by sliding the wing back and forth on the wire, force the wire thru the body until it emerges on the opposite side. Then grasp the wire and pull it through to a length just one and one-half the width of the body at that point. Bend the wire over into a half-loop and clinch it back into the body

by pulling it from the opposite side. Repeat this clinching operation with the other wing wire. Now insert the leg wires in a similar
fashion, clinching both in the body as shown in Figure 8. The same
procedure is followed if using a balsa wood body. In that case small
holes are drilled thru the body to receive the wires.

Make sure that the junction points of both wings and legs are at
the correct locations. Refer to the sketches made earlier. It should
be remembered that the junction points of the wings are generally
near the center of the body and that the legs are usually joined to
the body well back and at points slightly higher than the wings.
Sketches showing these vital junction points will now prove their
value because the final shape of the mounting depends largely on
accuracy in following these focal points.

Lay the bird next on its back and bring the skin together along
the incision; with needle and well-waxed thread, sew up the incision,
starting at the breast. The best stitch to use is one in which you
always insert the needle point underneath the skin; that is sew from
the inside, out. Take a few stitches loosely, then pick out any
feathers caught and draw the thread up snugly.

When the incision has been sewed up, bend the legs and wings
roughly to shape and insert the leg wires into a drilled block or
perch that will serve as a temporary support. Now prepare a

Bird-tail support wire.

sharpened wire bent into a T-shape at one end, and push it into the
rear of the body, just under the tail quills. (See accompanying
sketch.) This provides a support for the tail which can be adjusted
as desired. Pose the bird as naturally as possible, adjusting the
head, legs and tail until the desired attitude is attained.

If the wings are to be at rest and were not wired, pin them to
the body in their natural position with sharpened wires, each bent
into a small hook at one end. One or two such pins, strategically
placed in each wing, will suffice to hold the wings in place.

Insert a ball of cotton into the bird's mouth and tie its bill shut
by running the needle and thread through the nostril openings and

tying the thread underneath. The mouth may be left open slightly in some mounts. In such case, fill the mouth cavity with papier-mache and add a tongue carved from wood and properly colored.

Cut the wires from the artificial eyes and mount them in their sockets with clay or papier-mache. Use the needle to adjust the

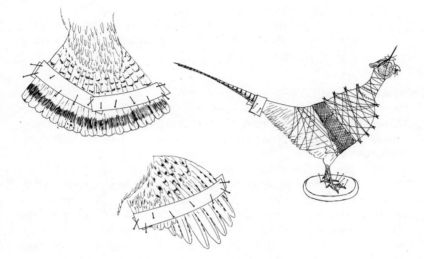

Figure 9. Finished bird wrapped for drying. If wings and tail are spread, hold quills in place by cardboard strips as shown.

lids evenly around them. Be sure the eyes are set in perfect symmetry, their correct positions found after observing them from all angles. They should protrude slightly, but should not bulge.

Cut two strips of light cardboard about one-inch wide for most birds and narrower for small ones, but of a length sufficient to reach across the tail. Fan the tail quills out to the desired width and lay one cardboard strip on top and the other underneath, pinning them together by pushing common straight pins through both pieces from above. On wing-spread jobs, cut similar strips for each wing and pin the flight feathers. (See Figure 9.)

Preen the feathers over the whole skin, making sure that all are in proper place, each fitting snugly over the next.

Finally, cut 6 or 8 thin wires of convenient length, sharpened at one end and bent into a U-shape at the other, and insert them in a row along the centerline of the bird's back and breast. Now, lightly wrap the plumage in place with turns of soft cotton cord round and round the body, using the pins to hold the windings in place. (See Figure 9.)

When the bird is completely dry, a condition reached usually after a week or more, the plumage windings may be cut away and the pins withdrawn. Again preen the bird where necessary with a small tweezers or needle, and remove the cards from wings and tail.

The bird is now fastened to a permanent base and all protruding wires nipped off underneath the feather surfaces. Apply varnish or color only where required.

MOUNTING LARGER BIRDS

Comparatively few amateurs have occasion to mount the largest varieties of birds. These are generally handled in the following different manner.

Only the feathered skin is stripped from the body. Thigh skins are slit down on the inside to the knee joint and severed there, allowing the legs to remain attached to the carcass. The head skin is inverted over the skull and detached from the bill, allowing the skull to remain attached to the body. The skin is then cleaned up, wings and tail cleansed of remaining fat and tissue and preserved with borax, and the whole laid aside in a dampened condition while the artificial body and skeleton is being prepared.

Preparing the Form

After determining and recording measurements of the bird, the flesh and entrails are removed, reducing the body to a skeleton without cutting tendons and ligaments holding the bony frame together. The skeleton is then thoroughly washed and preserved for use in supporting the mount. A block of balsa wood is shaped to fit within the bone structure of the body so as to provide an anchorage for supporting the leg wires. A similar wire is run through the skull and neck vertebrae to the block.

Now layers of papier-mache are built up on the assembly until a form has been modeled approximating the size and shape of the bird. (Techniques for making forms of papier-mache are explained in Chapter 2.) After these have completely dried, the form is waterproofed with several coats of shellac. The preserved skin, having been kept pliable, is now pulled in place over the form, carefully adjusted, and sewed up. The skin, wings, tail, etc., are filled out where necessary and pinned in place, wrapped, and except perhaps for wiring, otherwise handled as with smaller bird mounts.

As in all forms of taxidermy, variations in methods and procedure are employed by enterprising taxidermists to suit individual requirements.

Wiring the Wings

The wings of all large birds and those having spread-wing attitudes must be adequately wired to support them properly. This is generally accomplished by running sharpened wires through the tips of the wings along the wing bones and anchoring them in the artificial body or center block. The wires are securely fastened at intervals to the wing bone structure with individual ties. When thus anchored and supported, the wings may be bent at the joints to any natural conformation, especially when flight feathers are positioned and held in place by pinning and carding with cardboard strips.

Fleshing the Wings and Legs

Many amateurs (and professionals too!) neglect important fleshing of skull, tail, leg and wing parts. Inadequate cleansing of fatty tissue from these parts invites decomposition resulting in insect infestation, feather slip, and other troubles. Matter removed at these points should be replaced by padding carefully wrapped in place and neatly sewn up. Marrow should be removed from retained leg and wing bones of birds of any appreciable size. Drill holes in their ends and remove as much of the greasy substance as possible with cotton-tipped wires or common pipe cleaners. Remember that the less tissue and grease allowed to remain on the skins, the greater the chances of success. This is especially true when the borax alone is used as a preservative. Conscientious effort in these matters pays off in producing quality work.

HAWKS AND OWLS

Many birds of prey have protruding cartilaginous bones above the eyes. These should be retained on the skull since they provide the facial expressions peculiar to these birds. Particular care must be taken in skinning the heads of owls because of their large and deeply set ear openings. Detach the skin deeply within the skulls of all birds. Owls also have a cup-shaped bone structure surrounding the eyes which must be left in place for proper facial expression.

It will be found that bird skins will stick most tightly to the body at the small of the back, just forward of the tail. Work carefully at this point, cutting closely to the bone. American woodcock and similar types are especially difficult to skin because of this. (Woodcock, incidentally, are among the most difficult of all birds to mount.)

Also, remember that if trouble is experienced in slipping the head skin over the skull on large-headed birds, an incision at the nape of the neck can be made through which the head can be inverted.

RINGNECK PHEASANTS

Ringneck pheasants are among the most commonly-mounted birds. Of convenient size for mounting and tough-skinned, they are

comparatively easy to mount. Amateurs commonly fail to spread the crimson wattles by inserting extra filling through the eye and mouth openings. Fan out the ear tufts and hold them in place with soft cotton cord while drying. A blunt wire rod moved about through the eye openings causes the head and neck feathers to fall neatly into place, producing an incredibly beautiful iridescence.

Neck skins can be inverted over the skull without nape incisions in ringnecks, although this is about the limit based on head size. Pheasants take a special multi-color eye about 11 m/m in diameter.

Since these birds are normally on the ground and seldom perch, they are generally displayed on a flat-type base. They are also adaptable, however, to wall panels and they make fetching trophies in dead game, running, flying or other novel attitudes. See sketch.

USING COMMERCIAL OR HOME-PRODUCED BIRD FORMS

Many up-to-date taxidermists use prefabricated body forms which have become available in recent years from various supply dealers. These are accurately molded to exact physical proportion and are made of compressed paper, cork, plastics and other modern materials. They can be had in various sizes for most of the common game species, waterfowl, and birds of prey.

Skinning and mounting procedures differ according to the type of artificial form used. When the head and legs are included in the

Examples of artificial bird forms made of plastic, cork and balsa wood.

form, skinning is done through an incision made down the back of the bird, instead of by the conventional breast cut. Leg bones are severed at the knee joint and the skull is cut from the bill. The skin and wing bones are prepared for mounting in the usual manner. The neck skin is then inverted and the head of the form fitted in place with its protruding wire seated in the upper mandible. The neck and leg parts of the form are flexed together, permitting insertion of the leg wires into the skin, down the backs of the legs and out the bottoms of the feet.

The leg skins are then carefully drawn up and tucked into a slot cut into the form at leg-body junction point. The job is then straightened up, stitched, and posed as desired. If the wings were wired for spreading, the wires are back-hooked through both sides in the usual way, or clinched to form-wires provided for the purpose.

CHAPTER 5

Mounting Small Animals

THE SKINS AND HIDES of all mammals except the smallest varieties must be properly tanned before mounting. It is, however, entirely practical for amateurs to make raw skin mounts of small mammals up to the size of fox when the wiring and packing methods here described are used.

Skinning small animals is usually work accomplished rather easily since most such animals are tough and durable, resisting accidental cuts. Special care must be taken, however, at critical head and face areas. A lot of taxidermists do a first-rate job in all but face details. The head, of course, is the focal point in any mount and deserves the most attention.

A major fault too often found in amateur work is making body forms too large. It is far better to err in the other direction because a certain amount of skin shrinkage will offset this mistake. Keep this in mind, therefore, when form-building and filling operations are under way. Use care in fashioning armature wires used in mounting because they have a disconcerting way of working to the surface where they show under the skin, especially if made too long.

Taxidermists should always use good judgment in matters of

health and safety, avoiding specimens that are unwholesome to handle in any way. Use care also in handling edged tools.

Before starting the skinning operation, it is well to jot down a few notes and sketches as to the physical proportions of the specimen. Take measurements from the point of the nose to the corner of the eyes; from the point of the nose to the back of the skull, and from the point of the nose to the root of the tail. Also measure the length of the tail from its root to the end of the vertebrae (not to the end of the fur or hair).

Measurements of the animal's circumference at its chest and abdomen will be helpful. Outline sketches of the limbs and the head will be of great aid, too, when final mounting is done, or when ready-made forms are to be made or ordered. Becoming familiar with the general contour of the mammal in all its various natural attitudes is also of the highest importance.

The following detailed instructions, for making raw skin mounts of small mammals offer easy procedures by which the amateur can obtain good results with a minimum of experience, even if using simple tools and materials. After several successes with these primary methods, the reader may wish to advance to more sophisticated procedures employing prefabricated body forms, as outlined later.

FIVE STEPS

It is possible to regard the task of mounting a small animal as involving five distinctive phases or steps, some of which are generally comparable with the steps for bird mounting.

Step 1 - Skinning

Place the mammal on its back, head to left, and proceed with the skinning as follows:

Make an incision starting at the base of the neck and running down the center line to the vent. Be careful to cut no deeper than necessary to sever the skin, especially at the abdomen. Grasp the edges of the skin on each side of the incision between the thumb and forefinger of each hand and pull the skin away from the body. With this start, it should be an easy matter to pull the skin free with the fingers, all around to the back of the mammal. If the skin is too difficult to be freed from the body with the fingers alone, use the knife, cutting with careful, sweeping strokes against the skin. The amateur will soon get the knack of this and before long will be wielding the knife swiftly and dexterously, making each stroke tell.

Work rearward from the abdomen until the junction points of the back legs are reached, then loosen the skin as far as possible around the thighs. By grasping the hind leg of the mammal and bending the knee sharply upward, the skin can be made to slip over the knee joint. It is then possible to work the skin entirely free around the lower leg. Sever the skinned leg at the hip joint and finish skinning the leg by stripping it down to the ankle. Repeat this operation with the other leg. If the mammal is larger than a squirrel, retain the leg bones to the hip. Clean all the flesh from the bones, but do not disjoint them. If the mammal is the size of a squirrel or smaller, the leg bones need not be retained, and the skinned legs can be severed at the ankle joint.

Now free the skin from the back end of the body until the root of the tail is reached. If the animal has a thick, fleshy tail, the skin should be split on its underside almost to its tip and the tail bone removed. Most common mammals have slender, bony tails such as that of the squirrel, the fox, etc., and these need not be split, but the skin can be slipped off between the thumb and forefinger. If it proves too difficult to remove the tail bone with just the fingers, clamp the root of the tail in a vise and with the help of two round sticks held in both hands, the most stubborn tail skins can be stripped. (In some cases, notably the porcupine, the tail must be literally carved out of the skin.)

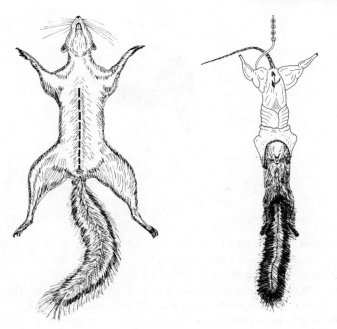

Figure 1. Opening incision.   Figure 2.

The partly-skinned mammal will now appear as in Figure 2. Now impale the body on a strong wire hook and suspend it from a rafter or other overhead point at a convenient working height. Pull the skin downward, inverting it over the head as in peeling off a glove. Use the knife deftly, cutting on all sides as the skin is brought downward until stopped by the front legs. The latter are encircled with the fingers until the skin is freed to the wrists in the same manner as the back legs. Here, again, retain the bones to the shoulder in the larger mammals, but sever the entire leg at the wrist in mammals the size of a squirrel or smaller.

When the skin has been drawn down to the skull, watch for the ear entrances. Extreme care must be taken here to cut close to the skull bone. Cut through the ear cartilage of the ear canal, as close to the skull bone as possible.

After passing the ears, the eyes are reached, and again extreme care must be taken to cut carefully, close to the bone. Stretch the skin over this point and cut forward until the transparent membrane covering the eye appears under the knife. One way to avoid mistakes is to place the fingertip against the eyeball from without and then

Figure 3.

cut against it until all the eyelid is freed from the eyeball without having cut the delicate skin surrounding it.

Continue skinning down the front of the head to the nose, cutting the cartilage of the nose close to the bone, and cutting the lips away from the skull at the gum line, all around the mouth.

The skin is now completely freed and will appear as in Figure 3.

Step 2 - Preparing The Skin

Assuming that all the flesh has been scraped from the retained leg bones, attention must now be given to the paws or feet. They should be incised on their bottoms and as much flesh as possible removed from the foot and hand bones except on the very smallest specimens. Following this, they should be rubbed thoroughly with powdered borax, packed to their original size with papier-mache and sewn up. Be sure to wax the sewing thread thoroughly.

Now go over the flesh side of the skin with the scraper, removing all evidences of adhering fat and bits of flesh. Some mammals such as raccoon require more efforts in this respect; in any event, time spent on this chore will be well repaid in the final mounting. Give special attention to the head portion of all skins. All adhering flesh must be shaved from areas around the eyes, nose and lips. Lips must be split to their edges and thinned out. The more lip skin retained, the better, as this allows better control of mouth details later. With extreme care and judicious use of the knife, the nostril linings can be freed from their cartilaginous fastenings and the nose thinned to a flexible condition. This provides for easier nose modeling later and will prevent the unnatural, wrinkled appearance to be noticed in many poorly mounted specimens.

Ears, too, must receive attention, the butts cleansed of excess cartilage, and all flesh removed. Where possible, and this is important, the ears should be skinned to their tips on all mammals. This is generally an easy proposition, but it can be troublesome if the ears are dried out. The skin on the back of the ears is separated from cartilage lining to the tips. A piece of heavy-gauge wire flattened and rounded at one end becomes a helpful tool for this job. Of course, care must be taken to avoid tearing the thin ear skins.

Now clean the fur side of the skin, removing any blood stains, grease spots, etc., with a weak solution of cold ammonia water. Warm water with detergent or a little borax added is a good cleanser for very dirty skins and even soaking the entire skin in a saturated borax-water solution can be recommended. In fact, such soaking, together with subsequent fluffing of furs with dry powdered borax, pays dividends later because a clean skin thus handled will resist deterioration for a much longer time. Oily skins should be degreased with naphtha or gasoline, if at all indicated.

Rub borax in thoroughly, on all parts of the skin, on bones, ears, nostrils, etc.

At this point, prepare a set of ear liners cut to fit the ear skins perfectly. Use lightweight fiber or high-grade cardboard for this

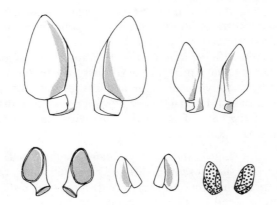

Examples of ear liners, moose, deer, fox, etc.; purchased and/or homemade of paper, Celastic, sheet lead, etc.

purpose. These should be dipped in hot wax or painted with shellac for waterproofing. Thin sheet lead, if available, may also be used if the ears are large. Don't forget to insert the ear liners before placing the head skin over the artificial skull. They should be pasted in place, if large, and secured by several stitches with needle and thread, the stitches being removed after drying. Work a small amount of thin mache or heated wax into the space between the ear cartilage and the backs of the smaller ear skins if no liners are used. This will stiffen them and prevent their shrinking and curling while drying.

When all this has been accomplished, the skin is ready for wiring and final mounting.

Step 3 - Preparing the Skull

There are a number of methods of preparing the artificial skull, several of which will be discussed here.

Sever the skull from the neck, close to the base of the bony structure and lay it on a sheet of paper and, with a pencil held vertically, trace its outline as seen from above. Then hold it on its side and trace its contours, as viewed from the side. Indicate on the sketch the location of eyes, jaws, teeth, etc., and make an accurate picture for use in preparing the artificial skull.

Some taxidermists use the original bony structure of the skull, rebuilding muscles and tissue removed by the use of papier-mache, while others prefer a hand-carved wooden facsimile. Still another method is to make a plaster mold, using the original skull and preparing a form of laminated paper. Techniques for making the latter are described in Chapter 2.

If the animal being mounted is the size of a squirrel or smaller, the amateur should proceed as follows: Clean the original skull thoroughly, removing all meat and tissue, eyeballs, tongue, and brain. Remove the brain with a small wire hook through the occipital opening at the base of the skull. This work can be facilitated by parboiling the skull in water.

When the skull is thoroughly cleaned, rub it well with borax. Now, replace the muscles and tissue removed by building up with modeling clay or papier-mache. Round off the skull until it resembles the sketches made earlier. It can now be laid aside while proceeding with the next operation.

If the animal being mounted is larger than a squirrel, it is recommended that an artificial skull be carved as close as possible to the contours of the original from a block of some soft wood such as white pine or balsa. Balsa wood has the advantage of being light and easy to handle. With the aid of sharp knives and wood rasps, this carving need not be difficult.

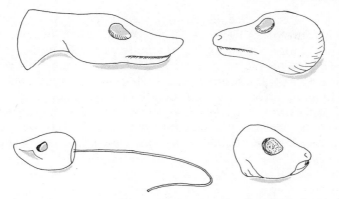

Examples of artificial skulls—purchased and/or homemade, wood-carved or of paper, plaster, etc.

Be particularly careful when carving the lips, undercutting the upper lip and forming a groove or lip pocket all around the mouth into which the lip skin may be tucked and pinned.

Sand the carving when it's finished and give it several coats of shellac.

Step 4 - Making the Wiring Assembly

In Step 1 it was explained that it was unnecessary to retain the leg bones on the smaller mammals. Assuming that the work is being done on such a mammal, proceed as follows: Cut two pieces of wire of correct weight and two and one-half times the length of one front leg, from foot to shoulder joint. Cut two more pieces two and one-half times the length of one back leg, from foot to hip joint. These are the leg wires and each should be sharpened at one end to a smooth tapering point.

Now cut another piece of the same wire one and one-half times the length of the body, from the tip of the nose to the root of the tail. This will be the body wire and it is also sharpened at one end. Cut a sixth wire one and one-half times the length of the tail, and sharpen it.

Stretch the prepared skin, fur side out, on the work table and insert the prepared skull into the head skin. Take the sharpened end of the body wire and force the point through the base of the skull (slightly above the occipital opening) and push it through the skull until it emerges at the nasal cavity. Let the end of the wire protrude through one of the nostrils of the skin for several inches. Now, at a point where the shoulders begin, whip the body wire into a small loop. (See Figure 4.) Make this loop just large enough to admit two of the leg wires. Then, at a point just forward of the hip line, turn a similar loop in the body wire. Clinch the wire to itself with a turn or two, then cut off the remainder, if any. (See Figure 4.) Now insert one of the foreleg wires into the bottom of the foot and force it through the foot until it appears within the skin. Pull it through the upper loop of the body wire a short distance. Repeat this wiring operation on the other leg. Then twist the ends of the two leg wires tightly to the body wire as shown in Figure 4.

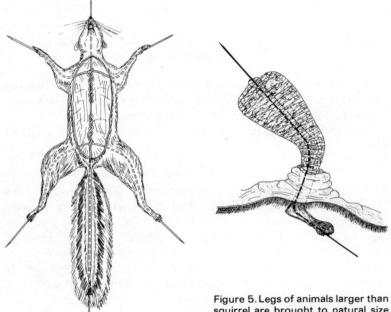

Figure 4. Wiring details—animal.

Figure 5. Legs of animals larger than squirrel are brought to natural size and shape with cord-wrapped excelsior and cotton as shown.

The artificial tail is the next consideration. Spin a quantity of cotton batting on the tail wire, wrapped smoothly in place with cord. Refer to your measurements and sketches. Make sure the artificial tail is just slightly smaller than the original, tapering it evenly toward its sharpened end. Moisten the artificial tail with water, then rub it well with borax. Now insert it into the tail skin until the sharpened end is forced through the tip. Feed the butt end of the tail wire through the lower body wire loop and wrap the excess around the body wire. Now wire the rear legs in the same fashion as the front, crossing them through the rear loop and twisting each tightly to the body wire.

In mounting mammals larger than squirrel (those in which the leg bones were retained), the wiring is accomplished in much the same manner except that the leg muscles are built up during the wiring operation. (See Figure 5.) When the leg wire has been brought through the foot to a point just beyond the upper leg bone, bind the bones loosely to the wire at several places. Then build up the artificial muscular structure of the leg by wrapping the wire-bone foundation with cotton batting and excelsior, binding some with turns of cord around the bone-wire assembly. Refer to the sketches made earlier to get these properly shaped. When the leg has been built up smoothly to the proper contours, the leg wire is pulled through the body wire loops and twisted thereon as already explained. In wiring up the larger jobs such as coon or fox, a more rigid assembly may be obtained by whipping the body wire into a pelvis-like structure at the back end, as shown in Figure 6.

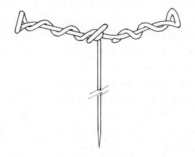

Figure 6. Alternative pelvic wiring for larger animals. Twisting body wire into pelvic-like shape as shown makes for a stronger job on larger mountings.

Step 5 - Filling and Final Assembly

Again, assuming that the mammal being mounted is of the smaller size, proceed with the final mounting as follows: Tamp small wads of cotton batting lightly in the neck skin, close to the skull. Be sure to pack the cotton evenly on all sides of the body wire, as it is important that this wire remain centered in the mount and that it does not later come to the surface just beneath the skin where it would show. Continue filling the neck skin as evenly as possible until it is filled to the shoulders. Then with the aid of a small stuffing rod made from a piece of scrap wire, pack the front leg skins using small wads of the batting at first, where the lower parts are thin and narrow, and increasing the size of the wads where the legs widen out. It is important that this packing be done in a careful, efficient manner, using just the right amount of filling, evenly packed around the leg wire.

Avoid getting a lumpy effect by pushing the filling around within while squeezing from the outside with the fingers. The legs may be packed rather tightly, but be careful not to overfill. In fact, overfilling is probably the greatest mistake made by most amateurs.

When the forequarters have been filled satisfactorily, do the same with the rear legs, using the same procedure. Now start sewing up the incision with needle and well-waxed thread. Take a few stitches loosely, then pick out any fur or hair that may be caught by the thread and draw the stitches up snugly. The best stitch to use is that in which the point of needle is always inserted from the under, or flesh side; i.e., from inside out. Sew the incision part way down, starting at the head end. Then continue the filling operation, packing the body skin evenly on all sides. Remember to keep the body wire well centered and equally surrounded on all sides with the packing. A good way to preclude the body wire from later working through the packing is to lay a broad strip of batting directly under the body wire and along its entire length. Continue sewing and packing alternately until the task is completed.

When the operation has been completed, pose the mammal in the desired attitude. Here is where a natural taxidermist's aptitude shows up, for the real test in mounting skill lies in the taxidermist's ability to mold his work into natural form configurations. Fasten the mount to a temporary base board drilled to receive the leg wires. Go over the body carefully with the fingers, smoothing out any irregularities, lumps, and hollows that should be corrected. With the thumb and forefinger many of these defects can be adjusted just by pressing and squeezing. If there are hollows that cannot be corrected by pressure on either side, take a sharp awl or upholsterer's regulator (simply an extra-large needle), forcing the point through the skin and picking the filling underneath until it fills out the depression.

Now turn your attention to the head. Adjust the head skin carefully on the skull, paying particular attention to the ears and eyes. Fill out the eye openings with mache or plastiline clay and insert the glass eyes, adjusting the lids with a needle until a lifelike expression is attained. Be sure the eyes are set at the correct depth, protruding slightly, but not having a bulging appearance. Remember, some specimens are naturally bug-eyed, while some have eyes slanted to mere slits! In any event, all must be perfectly symmetrical.

Insert a small quantity of mache into the ear butts from the outside and press the ears into shape with the fingers. A pin or two in the correct spots will help hold the ears in place.

Finally, adjust the nose and mouth details. A little mache strategically placed in the nostrils facilitates the smooth modeling of this important point. Fill a little material in the lips, if necessary, adjusting the lower lip first. If a wooden skull was used and a lip groove provided as was suggested, the work is much simplified. Tuck the lip skin into the groove evenly, all around its circumference. Secure the lips with common straight pins, using as many as necessary to hold the lips in place. A few well placed stitches with needle and thread in the mouths of small rodents such as muskrats, squirrels, etc., will hold the lips in place in cases where exposed teeth are desired.

The jaws of carnivorous (meat-eating) animals are seldom used in taxidermy because of the tendency of natural teeth to crack and discolor in time.

The mount may now be put away in an airy place to dry. Check the specimen occasionally while it is drying, and correct any distortions which may occur.

When the mount is completely dry, it may be removed from its temporary base and secured to a permanent one. Remove the pins and cut the excess wire protruding from skull and tail. Comb the fur and apply color to claws, nose, lips and eyelids where necessary. Breadcrust, passed lightly over the hair, imparts a pleasing lustre.

It is suggested that many of our native game animals be mounted on wall panels in various natural attitudes; perhaps with rustic or handpainted backgrounds. These panels are hung on the wall where they will usually be beyond the reach of curious admirers. Too often the latter have a habit of wanting to handle the mountings, which is to be discouraged.

USING COMMERCIAL OR HOME-PRODUCED ANIMAL FORMS

Prefabricated animal forms made of compressed cork, laminated paper and various compositions are available for most of the commonly-mounted mammals. These accurately-molded manikins eliminate much work and assure excellent results. They also can

Examples of small mammal (or animal) body forms, purchased or homemade of flexible plastic, cork, carved balsa, etc.

be used as guides for homemade reproductions, carvings, paper modeling, etc.

When prefab forms are used in mounting, conventional skinning methods are sometimes varied to accommodate the form. Some forms require a top incision, whereas in some cases small mammals are case-skinned, that is slit from one hind foot to the tail, then down to the other foot and skinned forward over the head, as in removing a sock. No bones are retained and only the nails are left attached to the skin. The skin is then prepared in the usual manner, a small ball of papier-mache being placed at the base of each foot, and the whole pulled over the form. Sharpened wires are pushed thru the foot skins and anchored within, if the form lacks built-in wires.

The specimen is then modeled and pinned where required, combed and set aside to dry. The use of paste over the form is optional and depends on the judgment of the taxidermist.

CHAPTER 6

Trophy Head And Large Animal Mounting

COMPARATIVELY FEW amateurs have the inclination or opportunity to undertake the full-sized mounting of large mammals because of the considerable effort, expense and experience required. However, once interest has been whetted by successes in bird and small-mammal mounting, amateurs may look forward with anticipation to larger big-game work, "where the action is," so to speak! Remember that for these, however, appropriate display space is needed.

THE TROPHY AND ITS FIELD CARE

Preparations for mounting should start immediately, in the field, if possible. Specimens in a fresh and clean condition are much easier to mount.

Skinning should be done at the work table, if at all possible. If done elsewhere, select a clean surface, avoiding an accumulation of litter and dirt which can prove difficult to remove. Guides or others who may help in the work should have close supervision because many good specimens are spoiled by the inept handling of inexperienced persons. Scalps and hides should never be folded sharply on themselves, but rolled in such a way as to suffer a minimum of damage when packed or transported. Watch out for rope cuts, saddle rubs and the like, and always avoid pulling of the hair or fur. If possible, blood stains should be cleaned off with cold water while still fresh.

Hides and skins must not be exposed to the sun or other source of intense heat for any length of time. They deteriorate rapidly if kept rolled up too long, the thus generated heat rendering them worthless in a short time. Prompt and thorough salting is a must if skins must be held for any extended period before tanning. Use plenty, and rub it in on every part, especially on thick or fleshy areas. The hides of large animals are sometimes so thick at places that it may be necessary to make a series of criss-cross cuts on the fleshy side to insure penetration of the salt which is otherwise effective only to a depth of about three-eighths of an inch.

POINTERS FOR GOOD WORK

Since the head is the focal point of any mounting, and the wall head is probably the most popular of trophies, the following detailed instructions are primarily concerned with head mounting. Most of the procedures in whole body mounting are simply extensions of this work.

First of all, let us have an understanding that all large animal skins must be properly tanned before mounting. This means that the amateur must send the hide to a reputable tanner specializing in the work, or do the job himself. Since the leathering of the larger mammal skins and hides is a difficult and exacting task and beyond the scope of this chapter, professional tanning is suggested for the amateur. However, those interested in doing this work are referred to the text on tanning which offers the needed detailed information (Chapters 9 and 10).

We will also have to consider it as accepted fact that the only feasible methods for achieving good big-game mountings are those making use of prefabricated (commercial) forms over which the skins are mounted. Crude, old-fashioned methods of built-up natural skull and bone structures and raw skin mounts are simply out of the question. Modern prefab forms and excellent tanning services are widely available at comparatively reasonable cost, making the work much more pleasant and eliminating a lot of time-consuming drudgery.

Before describing the work procedure for mounting a big-game or trophy head, the most common troubles encountered will be reviewed so that they may be kept in mind and avoided or overcome as the work progresses.

Form Quality and Fit

The use of quality forms of correct size as determined by accurate measurements will eliminate trouble with fit. Dry scalps, as received from the taxidermy tanner, often look as though they could never be stretched to their original size and shape, but such problem is rarely the case. Soaking and careful stretching of all parts will bring them to a perfect fit. Remember, too, that some animals taken in or near the mating season tend to have swollen necks and require a thicker mounting form.

Lip, Eye and Nose Details

It may be found that the amount of lip skin in relation to the space provided in the lip groove of the form will cause trouble when one tries to shape these parts to appear naturally. Judicious paring and trimming of both skin and form, together with adequate pinning, usually overcomes this problem. The important thing here is to bring the lip skins together neatly and evenly, without exposing the inner linings.

Difficulty is often experienced by the taxidermist in achieving natural eye and nostril alignment because of insufficient anatomical knowledge of the animal being mounted. The needed knowledge is best gained from observation of live animals or study of a correctly-mounted specimen. Areas critical to a good-looking mount require using just the right amount of fill material at the right places.

Ear Preparation

Many taxidermists fail to recognize the importance of the ears in head mounting. Ears must be skinned on the back sides to their very tips and edges and fitted with liners that will shape them completely and naturally. Any cartilage or skin not bonded to the liner will distort unevenly, causing a ragged appearance. The base of the ear (butt), must be modeled carefully where it joins the head, according to the manner in which it is posed. Ear liners of various materials, some complete with molded butts, can be purchased for use in mounting or as a model for making others.

Symmetry

In final form, a mounted game head must look good when viewed from any angle. It must therefore be symmetrical and proportionately balanced in all its details. Even the slightest variation, on one side in contrast to the other, causes a noticeable distortion and detracts from the whole representation.

WORKING PROCEDURE

In addition to explaining the working procedure for mounting a big-game trophy, suggestions applicable to purchasing eyes and forms will be included. As in preparing to mount other trophy categories, however, some of the same preliminary steps will apply. Hence, the all-important topic of what measurements to make and what preliminary data to record offers a logical starting point.

Measurements to Make

Before the head is skinned, several measurements must be taken to learn the correct size and type of head form to be used in the mounting. Jot down the distance between the end of the nose and the corner of the eye as shown in accompanying sketch. Also, the

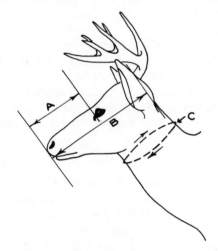

Measure game heads as shown to determine sizes of head forms to be made or ordered.

distance from the end of the nose to the back of the head, and the circumference of the neck just back of the ears. Measurements from brow tines of deer, etc., to nose tip will be of help in establishing the proper angle for fastening antlers. These measurements will determine the correct size of the head form to be made or ordered from the taxidermist supply dealer. It is also a good plan to decide at this time the type form desired, such as left or right turn, upright or "sneak," and whether or not the shoulders are to be included. It is good practice to make a simple sketch of the head, noting the position and angle of the antlers, the ears, etc., as well as all pertinent measurements. On a separate piece of paper make an accurate tracing of the ear flanges of all specimens. These will be of great help when ordering or making well-fitting ear liners.

Unless you intend only a head mount, accurate outline drawings of the animal body are necessary and all pertinent measurements must be recorded for use in determining the exact size of the artificial body form to be made or ordered. This is very important and must be done carefully and in much detail.

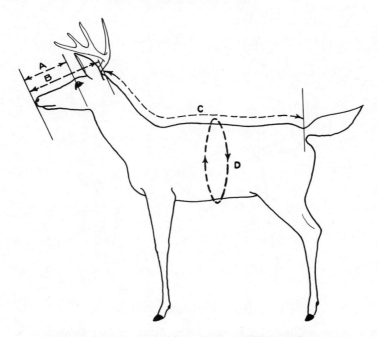

Measurements needed for ordering whole body forms.

It is important to specify the correct size needed when ordering forms from the dealer. As a rule, the measurements required in ordering forms are as follows: the distance, in inches, between the end of the nose and the inside corner of the eye; the straight line distance between the end of the nose and the back of the head, in inches; the length of the body from the end of the nose to the base of the tail, following the curve, in inches; the circumference of the body in the center. (See accompanying sketch.)

Skinning and Preparing the Scalp.

When the important measurement details have been taken care of, proceed with the skinning as follows: Make an incision length-wise along the top of the neck from the shoulders to a point just back of the antlers or horns. (In fur rug work, where the top of the head and neck are uppermost, the incision would be made on the under side.) At this point, continue the incision to the base of each antler in the form of a "Y" as illustrated in Figure 1. Then make

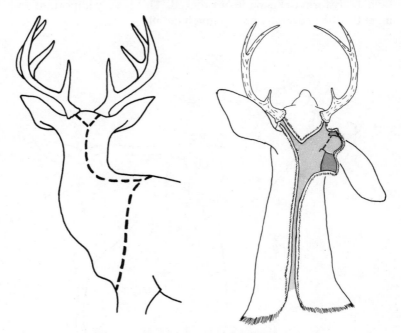

Figure 1. Opening incisions—game heads.

an incision around the neck (or the back of the shoulders, if a shoulder mount is planned), severing the neck or "cape" from the body hide.

Be sure to allow plenty of length to the cape, and under no circumstances cut the skin on the under or throat side. Make these incisions by forcing the point of the skinning knife through the hide with its sharp edge uppermost, and push in the direction desired. This will prevent any unnecessary cutting of hair. Now lay the head on the work table and skin the neck, working down from the lateral incision and forward. When the skin has been freed from the back portion of the neck, it may be inverted forward and held taut with one hand while cutting it free with the other. Use long sweeping strokes, as close to the hide as possible without cutting it.

When the ears are reached, skin them a short distance on the top side, then cut through the ear butts and cartilage, severing them close to the skull. It will be found that the hide clings closely to the bone at the base of the horns. Using a small knife or scalpel, cut the scalp carefully away from the bone just under the antler burr.

The next critical points reached will be the eyes. The best method of skinning around the eyes is to insert the fingers of the left hand into the eye socket and pull the skin away from the bone while cutting through with the right hand. By doing this, the operator can determine by sense of touch, the proper place to cut the membranes around the eye socket without damaging the outer eyelids. The skin should be cut very closely to the bone, especially where it is attached at the deeply set tear ducts, just forward of the eye corners. Use a sharp-pointed scalpel blade to cut the skin away at this point.

Now skin forward to the nostril and mouth openings. Again insert the fingers of the left hand in the mouth opening and cut the skin free from the lower jaw, using the fingers to guide the cuts and pulling the skin away from the bone. The lips should be cut from the jaw bones at the gum line, thus allowing plenty of lip skin to remain. This lip skin is used later to good advantage in molding mouth details. Cut the cartilage of the nostrils well back from the nose openings and keep it attached to the scalp.

The scalp is now laid aside and attention given to the horns or antlers. Use a meat or hacksaw to cut the skull in such a way as to leave the horns intact. Make the cut in a straight line, starting at the back curve of the skull and leading through the upper part of the eye socket as shown in Figure 2. Since head forms vary as to size of space allotted for fastening the antlers, it is best to allow plenty of bone to remain on the horns, as this can be cut back to fit when the head form is received. Clean all gristle and flesh from the bone holding the antlers and rub them with powdered borax, after which they can be laid aside until ready for fastening to the head form.

Figure 2. Cutting antlers from skull.

Now the scalp must be prepared before shipping the skin to the tanner. First, the ears must be skinned on their back sides and inverted over the cartilage as shown in Figure 3. This operation

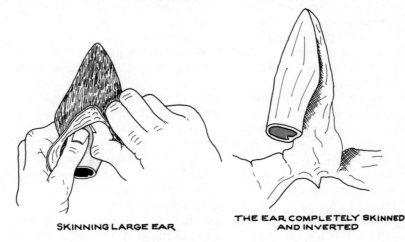

SKINNING LARGE EAR

THE EAR COMPLETELY SKINNED AND INVERTED

Figure 3. Skinning and inverting ears.

may cause some trouble for the amateur at first, but it is easily mastered. Insert the fingers into the ear from the outside and cut the skin from the ear butt cartilage with a small sharp knife or scalpel, until the butts are skinned, then with the fingers and thumbs of both hands, pull the skin free to the tips of the ears. (See sketch.) A dull rounded instrument such as the blunt end of a scalpel or a small, flat, hardwood paddle may be used in this to good advantage. It is important that this skinning be done completely, to the very edges of the ears at all points. There are special reverse-action plier tools available for this work. But avoid putting too much tension on the ear skins as they are thin and easily torn, especially at the edges.

Now trim and cut away the excess flesh around the eye openings, thinning the hide at this point, and carefully split the eyelid linings to their edges. Remove some of the lip tissue and shave away excess flesh, splitting the lips carefully to their edges at all points.

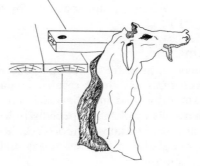

Figure 4. Slip the scalp over a simple bench beam to facilitate fleshing of nostrils and lips.

Insert a finger into one nostril from the outside and using this as a guide, carefully free the cartilage from the skin. With a little examination and study and judicious use of the scalpel, the nostrils may be separated and excess tissue removed from the nose parts, yet still retaining the delicate nostril linings. Avoid cutting through the skin at all times as too-deep cuts here mean a visible repair job later on.

When this work has been completed, go over the entire scalp, scraping away from it all remaining bits of flesh, fat and gristle using a sharp, broad-bladed knife. Spread the scalp out, flesh side up, and salt it thoroughly, rubbing plenty into all parts. Then fold the skin once, flesh-to-flesh, roll it up and place it on a sloping surface for a few hours to drain. Repeat this salting once more, and the scalp is ready for shipment to the tanner.

Proper salting "cures" the skin. It removes moisture, sets the hair and prevents spoilage. It will be found that cutting blades "take hold" better when removing flesh and gristle or thinning scalps at critical areas of eyes, nostrils, and lips after some of the moisture has been removed by the salt.

It should be remembered that the responsibility for adequate skinning, fleshing and salting to prevent spoilage lies with the taxidermist. The tanner providing the professional services, takes it from there, completing the work of cleaning, shaving and leathering the skins for mounting purposes. The tanners have the right mechanical equipment and the know-how to do the job properly. Information concerning commercial taxidermy tanners can be obtained from taxidermist supply dealers, advertisements in sporting magazines, or from Mac Rae's Bluebook and Thomas' Register of American Manufacturers, available in any public library where they are listed alphabetically by subject. (See also listings, Chapter 1.)

Purchasing Eyes, Forms, Etc.

A few words may be in order concerning the purchase of eyes, forms, and other supplies. Good results depend largely on the quality of workmanship and materials that are used in any job. Buy the best materials available, especially in forms and artificial eyes, as these are of the utmost importance. Reliable taxidermy supply dealers will gladly answer questions and recommend the best materials when asked.

Fastening Antlers

Assuming that the scalp, proper head form, eyes, ear liners, etc. are on hand, proceed as follows: First, fasten a strong board to the wooden backboard of the head form with several screws. Clamp this support in a vice or to the workbench in an upright position and at a convenient working height.

With a sharp knife or flat chisel, cut a narrow slit in the lip groove of the form, cutting completely through from corner to corner. Keep this opening very narrow. With the knife, round off the sharp edges slightly. The lip skin on the scalp will be neatly tucked into this opening. Now cut the skull plate supporting the horns to fit the notch in the head form. Drill several holes in the bone between the antlers, countersink them, and fasten the antlers firmly with three-inch brass wood screws; well-made forms have a block of wood built in to receive these screws. Then mix a sufficient amount of papier-mache to mold around the joint, making a smooth, rounded job as shown in Figure 5.

Setting Glass Eyes

Fit the glass eyes into the sockets, molding them in position with

Figure 5. Deer head form showing antlers and glass eyes set in place with papier-mache.

papier-mache. Hold them in place with a few straight pins around their edges until the mache is dry. Some taxidermists, however, prefer to set the eyes after the skin is in place, this method possibly permitting greater leeway in later adjustment. Nevertheless, it is important that the eyes be set to correct angle and depth, as their position is critical to the quality of the mount.

Scalp and Head Details

The leathered scalp (back from the tanner) must be soaked in a saturate solution of borax water made by adding enough powdered borax in cool water so that some remains undissolved in the bottom of the container after stirring. Make enough of the solution to cover the scalp completely and allow the scalp to soak for several hours. Move the scalp about occasionally, alternately squeezing and rinsing it until it is soft and pliable throughout. Now remove the scalp, squeeze out the excess liquid and stretch it in every part. Cut off the butts of the ear cartilage about a half-inch below the front opening and turn the ear skins back over the cartilage to their normal position. At this time it is well to stretch the ear skins carefully out to their fullest extent. When stretching and handling the scalp, never pull on the hair.

With a sharp scissors, trim off any ragged edges at the eyelids, lips and nostrils, but allow plenty of lip and nostril skin to remain. Go over the inside of the face and thin the skin uniformly at places where it is possible to do so. Now mend any cuts or torn places in the scalp, using fine stitches of strong waxed thread and a surgical needle of small size. Repair bullet holes, if any, as shown in Figure 6.

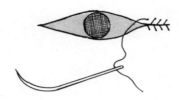

Figure 6. Repair bullet holes by cutting skin and sewing as shown.

Now give the scalp a preliminary fitting over the head form. Stretch the damp skin until it fits properly at all points and the neck seam closes neatly. Remove the scalp again and brush form paste into the ear pockets. Slip the ear liners in place, making sure they fit perfectly to the edges. Fiber, compressed paper and lead liners may be trimmed with a shears until a perfect fit is obtained. Now brush paste over the face and neck of the form and fit the scalp in place. Center it properly, then fasten it in position with a few temporary nails driven along the neck seam.

Sewing, Trimming, and Adjustments

Prepare a double linen sewing thread and large needle and sew up the neck seam. There are several ways to do this, most taxidermists starting at the top and working back as shown in Figure 7. Use the type stitch in which the needle point is always inserted from the under or flesh side. Another good method is to use a pair of needles, starting one at each antler and stitching to the junction and then lacing through holes previously punched at half-inch intervals along the neck seam, lacing it up in a similar manner as in lacing a shoe. Whatever the method used, the important thing is to get a tight even seam, drawing the hide snugly up under the antler burrs and catching as few hair as possible under the stitches. In some cases auxiliary nailing is required to keep the scalps from shrinking away from the bases of the horns or antlers. Use brass brads in holes pre-drilled in the bone for doing this.

When the sewing is completed, remove the screws from the supporting board and hold the head between the knees or other convenient position while trimming off the extra length of cape.

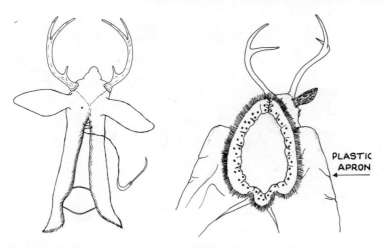

PLASTIC
APRON

Figure 7. Sewing up seam on partly-mounted head.

Figure 8. Showing method of securing skin to backboard. Hold the game head between the knees to nail the skin to the backboard as shown. This is the preferred method of doing this job because it precludes the dangerous possibility of the backboard pulling away from the form if the antlers or horns are very heavy as is sometimes the case.

The best method of fastening the skin in the back and one used by most taxidermists, is to allow enough skin to turn down over the baseboard where it is nailed neatly in place. The hair is shaved off and overlapping folds of hide are cut away to form a flush surface as shown in the sketch.

Another way to do this is to trim the skin evenly around the rear of the form so that it extends about a half-inch over the back edge. Cut through the skin from the flesh side so as not to cut any hair. Use brass nails to nail the skin in place. Spread the hair apart with the fingers of one hand while driving the nails. The nails should be spaced at three-quarter-inch intervals around the entire circumference. Make sure that the nails are long enough to penetrate the hide, the form shell, and well into the wood of the baseboard.

Finishing Up The Head

Now return the head to its former position on the supporting board and continue the mounting work. Push the neck skin smoothly against the form, squeezing out air pockets and bubbles. The skin may be moved about on the form to some extent by pushing it with a sharp instrument such as an upholsterer's regulator. Adjust the

Figure 9. Close-up views of eye setting. The angle and depth of eye settings largely determine the expression wanted in game head mountings; fear, curiosity, shyness, belligerence, etc.

skin at the eyes next. Arch the upper eyelid a little, just forward of center. Possibly a little mache applied at the proper places will enhance the character of the eye. (See Figure 9.) It pays to spend plenty of time adjusting at this point as a poorly set eye can spoil an otherwise good mounting. The eyelids may be held in position with a few strategically placed pins. Drive a small brass nail in both tear duct depressions.

Press a small amount of papier-mache around the mouth opening of the form and adjust the lower jaw and lips first. Tuck the skin into the lip slot with a dull instrument such as a scalpel handle or screwdriver. It may be necessary to lengthen or widen the lip slot at the corners of the mouth to accommodate the heavier lip skin at these points. Don't forget to brush plenty of form paste on the form before pressing the skin into position. Now apply a thin coating of papier-mache into the nostril depressions of the form and across the nose. Do not use too much modeling here because all good head forms have good mouth and nose proportions and require little or no modeling.

Brush a little more paste over the nose before modeling the nose skin in place. Tool a little mache into the nostrils from the outside and work both the upper and lower lips neatly into the lip groove. See that the nose and mouth details are perfect. While little or no pinning of lips should be necessary, several pins will be of help in holding the lip skin in place until dry. Later, they can be driven deeply into the mouth groove or removed.

Modeling tools of various shapes are valuable in forming face details correctly. When this work is completed, turn your attention to the ears. Using thumb and fingers, press the ear skins into firm contact with the liners and shape them to the proper contour. Then, with needle and thread, sew through the entire ear assembly, taking several long stitches and knotting the thread when it is drawn up tightly. These stitches will be removed when the mount is completely dry.

Many methods are used to hold ear skins in place. Some taxidermists use pieces of cardboard, fiber, celastic or other like material, shaped and fitted to front and back of the ear and held in place with wire staples. These covers can also be clamped in place with a row of common spring-type clothespins. Others prepare a set of carved wood inserts which are placed within the ear flanges. The ears are then kept in tight contact with the inserts and liners

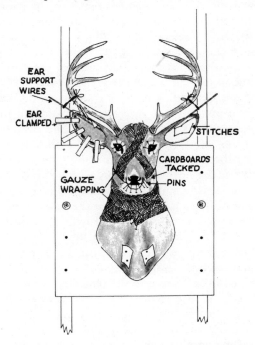

Completed game head prepared for drying. Note adjustable baseboard bolted to 2″ x 4″ vertical studs, an ideal working base for any wall-hung taxidermy jobs. Also note method of clamping ears.

by clamping or bandaging. You have the option of using any of these methods. (See accompanying sketch.)

The ear butts must be shaped up with papier-mache inserted through the ear openings. Model them to conform with the chosen attitude of the ears, a most important aspect of the job. The ears can be further supported by tying them to sharpened wire rods driven solidly into the head form through the ear openings. Bend the ears and their wire supports to the desired position. Tap all stitching and sewing repairs flat with a light mallet.

Final Touches

When these steps have all been accomplished, go all over the head again, checking all details before setting it aside to dry. Step back and observe the mount from every direction to make sure that it looks natural from every angle. The job is now finished up by combing through the hair and slicking it down with a light sponging of borax water. Some taxidermists wrap portions of the head lightly with gauze bandage and tack small pieces of cardboard to the skin where it does not stick snugly to depressed areas of the form. Check the head occasionally while it is drying and correct any slight distortion that may occur about the face.

When the mounting is *completely* dry (and this may take considerable time, depending on atmospheric conditions), remove the wrappings, if any, and remove all pins from the lips, eyes, ears, etc. Withdraw the wire supports from the ears with a twisting motion or nip them off inside the ear if they prove too difficult to remove otherwise. Also, remove any clamps or stitching from the ears. Brush the head thoroughly with a stiff bristle brush, especially the long hair in and about the ears. Then rub the head down with a clean, soft cloth, bringing out the natural lustre of the hair.

Mix a drop or two of painting oil with a small amount of tube oil pigment. Use a small flat-type artist's bristle brush to apply the oil paint to the bare skin of the nose, in the nostrils, and around the lips and eyelids. Carry the color in a narrow line from the corners of the eyes down into the tear duct depressions, as shown in illustrations. If bare spots show inside the ear flanges, tint these carefully with flesh color. If the dusky markings on the nose and jaws are not symmetrical, a little thinned oil paint of matching color, carefully applied, will often improve the condition. A little white coloring matter applied in the front corners of the eyelids imparts a lifelike touch.

The head may now be fastened securely to an attractive shield or panel, completing the job.

WHOLE BODY MOUNTS

If the body is to be mounted, the skinning incisions should be made accordingly. The dorsal or top cut is made from the back of head lengthwise to the tip of the tail, and the skin peeled down to the feet. The only other cuts made are incisions at the wrists and ankles for fleshing. When this method is used, the skin is drawn up over the body form from beneath, thus requiring but one seam for stitching.

Perhaps the more commonly used method of skinning is that of incising the skin from a point between the front legs lengthwise to the anus and tail, along the underside, then from one foot cross-wise to the other foot, on the insides of the four legs. This results in a flat skin which is draped over the manikin and sewed in place. The heads of antlered animals are skinned through a separate incision on the back of the neck so that the skin can be removed over the horns. In either case, no bones are retained.

Sometimes the locations of skinning incisions are dictated by the desired viewing angle of the finished pose. They are made, of course, where the seams would be invisible.

After the incisions are made, the hide is carefully separated from the carcass and removed in one piece. Skins are then fleshed, salted, rolled up, and sent to a commercial taxidermy tanner to-gether with a letter of instruction.

Several methods of building up body forms are used, all re-quiring a good knowledge of the specimen's anatomy. Older methods made use of shaped wooden centerboards to which were fastened steel leg rods bent to conforming measurements. The artificial skull, made of papier-mache or a carved-wood facsimile, was anchored to a neck rod and fastened similarly to the center-board. This framework, or armature as it is called, was built up by padding and wrapping, then modeled out with a layer of mache.

Others formed a meshed hardware-cloth covering over a wire armature, thus establishing a base capable of supporting the modeling material. Muscles and other parts of the specimen's anatomy were modeled from measurements and the form com-pletely dried and waterproofed.

The use of ready-made forms of laminated paper greatly simpli-fies this work, saves time, and produces a far superior job. Commercially-made forms are strong, light, and cast to exact natural proportions. Hardly anyone could possibly achieve such accurate representations by modeling at home from a set of measurements, however accurate.

After ear liners have been inserted, the prepared skin, having been soaked and stretched to size in all parts, is worked into position and pasted in place on the form. The seams are carefully sewed up and attention is next given to modeling the head and face details. As the mount dries, the hair is combed and brushed, the ears positioned, and nose and eye details checked at frequent intervals for distortion. Finally, the mount is transferred to its permanent base and coloring applied where necessary.

It would be most impractical, incidentally, for one person to attempt the body mounting of a big-game animal without the help of one or more assistants.

OPEN-MOUTH MOUNTINGS

Preparation of animal scalps for open-mouth mounting (wall head, fur rugs, etc.) entails just about the same procedure as outlined for game heads, regardless of size and type of specimen. Scalps should be tanned for a first-class job, although some taxidermists do mount the raw skins of some of the smaller animals such as fox and raccoon. While it is possible to build up natural skulls for use in head mounting, the time and effort required make it much more practical to purchase ready-made artificial forms, especially since these so consistently lead to superior mounting jobs. It is however, entirely practicable for the amateur to prepare his own head forms for the smaller, closed-mouth and half-head forms used in fur rug

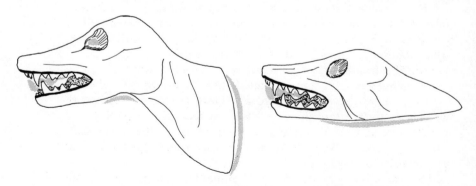

Examples of artificial skulls for open-mouth mountings.

manufacture and for the small-animal wall heads. These are easily carved from blocks of balsa or other soft woods.

Open-mouth head forms complete with teeth and tongues of rubber and composition are available for most of the animals generally considered for open-mouth mounting. Some dealers offer shells and jaws separately, but the home assembly of these often proves troublesome, therefore it is recommended that complete forms be ordered by the amateur. Fasten the lip skins of small scalps such as wildcat, fox, etc., to the inner edge of the form jaws with a row of very small, brass pins. (Dressmakers, "sequin pins" are excellent.) Then, when the skin has dried in place, model smoothly over them with papier-mache. Use larger brass pins or brads for the larger heads of wolf, bear, etc. Always use brass hardware in taxidermy work. Don't forget to first take measurements and make ear tracings before skinning an animal to be mounted as this cannot be done later. Ear liners must be purchased or made for all head mountings. Use leather, fiber, paper, celastic or sheet lead for homemade liners.

CHAPTER 7

Novelty Taxidermy

UNUSUAL AND NOVEL
ideas in taxidermy provide interesting departures from the
ordinary. There are unlimited possibilities in the field, possibilities bounded only by the imagination. Decorative and unusual
poses, group caricaturizations, impressionistic background scenes,
and the like are fun to improvise and can add much to the enjoyment of the hobby.

Animals like deer, elk and antelope provide the basis for useful
and attractive novelty possibilities. As is said about the pig, "Every-
thing can be used except the squeal." Here are a few relevant
suggestions and instructions.

MOUNTING ANTLERS AND HORNS

Deer and elk antlers make useful racks whether used singly or in
pairs. They can be cut off at the base and fitted with dowel screws
for fastening to wall panels. Use potassium permanganate crystals
dissolved in water to restore color to old bleached antlers found
in the woods. Stag-handle grips of all descriptions can be fashioned
from heavy antler material by sawing grinding, and filing.

To mount antlers as a trophy, remove the top portion of the

skull with a hack or meat saw, and at such an angle as will provide a flat base. Clean and trim the skull portion and rub it with powdered borax. Pull out the thin, tough layer of tissue lining the bone on the inside by gripping its edge with a pliers. Drill and countersink two holes in the skull plate and fasten it permanently with brass, flat-head screws to a suitable panel faced with waxed paper. A shaped-up wooden block may be placed between the bone and panel thereby providing a means for raising the antlers or tilting them as desired. Now mix a batch of rather thick papier-mache and model it neatly over the bone area. Finish by stippling the moist mache, then painting the base when dry. Finally, remove the wax paper by trimming it neatly around and away from the base with a sharp blade. Dry paints can be added to the mache while it is being prepared, if desired.

Another way to do this is to make use of the natural hair. In this case, simply retain the head skin, cutting it long enough to wrap under the bone base. Separate the skin from the skull except where

Mounting antlers on plaque.

it is attached at the base of the antlers. Rub borax on the clean, scraped bone, as described before and soak the assembly for a time in borax water. Now drill the screw holes from the underside and cut matching slits in the head skin. Secure the skin tightly around the underside of the bone base with crossing stitches. The job is completed by fastening it to the panel through the prepared holes. There are, of course, many other methods, materials, and variations of doing this work.

Steer, cow, buffalo horns and the like are boiled for several hours until the core can be removed. The horns are scraped lengthwise along the grain with files or rasps until dents, scratches and other imperfections are minimized. They are then buffed to a high polish on a motor-driven wheel using polishing compounds of emery, rottenstone, etc. A wooden center-dowel is shaped to fit snugly into the horn bases where they are fastened with screws. The work is completed by shaping up the centerpiece with papier-mache and applying padded velvet or tooled-leather covering, attractively set off with brass upholstery nails. Horns are traditionally used in the home manufacture of trumpets, powder horns, novelty trays, etc. The etching, carving or other art work done on these items is called "scrimshaw."

SMOOTH OVER WITH PAPER-MACHE, etc.

Mounting horns.

The horns of the Pronghorn (American antelope) require special attention. They are unique in that they are the only true *horns* that are deciduous (shed annually). The horns and skull plate are boiled in water until they can be removed from their cores. The soft tissue must be scraped down to the hard bony base of the core. Then the horns are cemented back in place on their bases with a soft mix of plaster of paris. If this procedure is not followed, the pronghorns will deteriorate in time.

ORDERING LEATHER AND SKIN NOVELTIES

Deer skins are an important source of fine leather, and are often termed "The Aristocrat of Leathers." As such, they should never be discarded or wasted. Commercially-tanned deer skins make attractive table or wall coverings whether in grain-on, suede-finish, or hair-on styles. Buckskin is traditionally American and tans into beautifully soft, supple leather for custom-made gloves, jackets, handbags and novelty goods, available in the latest styles and colors. The following table should aid in determining the leather yields of deer.

Dressed Deer Weights	Hide Size	Approximate Yield
90-130 lbs.	Small	6-8 sq. ft.
130-175 lbs.	Medium	9-11 sq. ft.
175-200 lbs.	Large	12-15 sq. ft.
Over 200 lbs.	Extra-Large	16-18 sq. ft.

One average-size deerskin yields enough leather for
   3 pairs of gloves.
Moccasins require about 3 sq. ft.
Handbags require 4 to 6 sq. ft.
Jackets and coats require 30 to 40 sq. ft.
Purses, wallets, key cases, etc., are made from
   remaining scrap leather.

You *do* get your own leather back from commercial tanners!
Mark skins by punching an identifying pattern on each with a sharp
awl ground to triangular shape providing for three-cornered
cutting edges. Mark the skins in an area where it will not interfere
with the manufacture of the goods. Remove all flesh from the
hides, then mark, salt and drain them thoroughly before shipping
them to the manufacturer. Most concerns allow discounts to taxidermist customers.

MAKING BIG-GAME FOOT NOVELTIES

Game animal feet are easy to preserve and mount. They can be
fashioned into a wide variety of articles both useful and decorative such as ashtrays, lamp bases, gun racks, book ends, and
thermometers, to mention but a few.

Raw deer feet are usually rather easy to come by, too, since
most of them are discarded by sportsmen. Check big-game hunting
friends or relatives before the season. Local butchers and frozen-
food locker people who process the animals for hunters can usually
be prevailed upon to save feet when asked and, more than likely,
the amateur taxidermist will wind up with more than he can use!

The legs are severed just below the knee and hock joints. Slit
the skin down the back where the hair is the longest, cutting from
the inside and thus avoiding cutting any hair. Pull the hide down
over the knuckle, nip off the dew claw attachments on the inside,
and with the knife or scalpel, skin the foot *completely* to the point
where it is attached to the hoof. Now grasp the large tendon and
pull it from the shin bone, the latter, however, being retained.
The tendon branches into two parts below the knuckle and is cut
off at the points of attachment in back of the toes. Remove most
of the cartilage until the knuckle is exposed, but do not disjoint
this hinge. Be sure to remove the marrow from the shin bone with
a long wire scoop. Scrape and clean all flesh and cartilage away
from the bone and skin parts. Neatly cut away the small, evil-

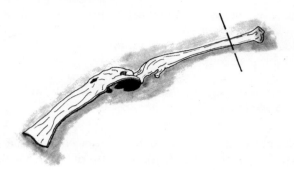

Section view of deer foot bone in elementary process.

smelling sac found in the skin between the toes. This opening will be sewed together before mounting.

The bone-in-foot skins are soaked for several days in a pickling solution made and used as follows:

Water - 1 gallon
Salt - 1 quart
Sulphuric Acid - 1 ounce

Bring the water to the boiling point to dissolve the salt.

Allow it to cool before stirring in the acid.

Use one pint measure of alum instead of the sulfuric acid if desired.

When cold, the solutions are ready for use. Crystal formations in the liquid do not harm its effectiveness.

Use a ceramic crock or wooden keg for this soaking, never metal.

Keep the sets of feet held together with plastic ribbons or nylon cord, and use homemade identification tags made of sheet lead, if needed.

Foot mounting is really a very simple process which can be done in several ways; perhaps the following suggested procedure will prove easiest for the amateur.

Rinse the leg skins in fresh water and stretch them to their fullest extent. Any remaining bits of gristle and tissue can now easily be scraped off. See sketch of tool, etc. Sew up the small opening caused by the removal of the scent gland between the toes. Beginning at the hoof end, stitch up the lengthwise incision using sturdy, well-waxed sewing twine. When partially sewn closed, press a rather thick mixture of papier-mache into the foot and on all

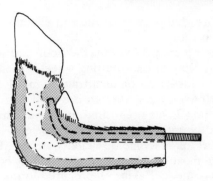

Section view of hand-processed deer foot for gun rack, clothing hook, etc., papier-mache filled, hand-shaped, using bone in process as suggested for amateur project.

sides of the bone. Pack this material in solidly, making sure that all spaces are filled. Then continue the stitching and filling until the desired length is reached. Mold the foot from the outside with the fingers as the work proceeds. Straight, upright foot jobs for use as lamp bases, thermometers, etc., require no additional inserts. Feet that are to be used as clothing hooks, gun racks and the like are turned up at 90 degrees with a bent reinforcing rod inserted in the back during the packing process. This strengthens the foot and provide a means of fastening the foot to panel or wall fixture.

The rods are easily made from long thin lag screws, or from machine bolts with threaded ends, their heads removed. Make

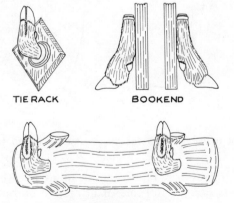

TIE RACK   BOOKEND

GUN or FISHING ROD RACK

Examples of novelties made from big-game feet. There are literally dozens of similar possibilities.

sure the threaded portion of the rod extends back to the butt of the finished foot.

When completely dried out, the hoofs are scraped, ground and polished by hand or on a buffing wheel. When ground down near the "quick," hoofs take on an attractive mottled texture and color. If rods with woodscrew threads are used, the fixtures can simply be turned into holes pre-drilled in a base panel. Panels must be drilled and countersunk on their reverse to accommodate nuts and washers of machine-threaded bolts, if the latter are used.

Actually most taxidermists remove the bone altogether and mount the tanned leg skins on prepared forms of plaster, wood, or compressed paper. These are available in a variety of shapes and sizes from taxidermy supply dealers and greatly simplify the work. The foot forms can be had with the necessary hardware for making lamp bases or other novelty items. (See illustrations.) Enterprising amateurs can easily reproduce forms of all kinds by means of plaster-casting methods, but this is largely unnecessary since

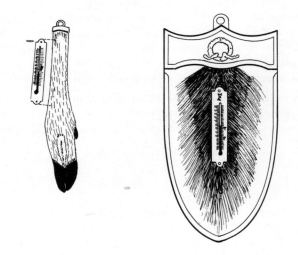

Big-game foot novelties: *(left)* foot thermometer; *(right)* tail thermometer.

commercial forms are relatively inexpensive. Most taxidermy supply dealers carry a full line of attractive chrome fittings and novelty hardware made to fit various big-game animal foot forms designed for the production of ashtrays, inkwells, thermometer scales, desk lighters, etc.

CHAPTER 8

Processing Skins For Collections Or Scientific Study

STUDENTS OR OTHERS seriously concerned with natural history study may legitimately wish to start or add to existing collections of specimens for scientific study and display. It should be pointed out that most wildlife is protected by law and may be taken only as prescribed by designated authorities. Furthermore, no reader is in any way encouraged to hunt or take any wildlife species simply for such purposes as mounting or "collecting." If in doubt as to either the legality or propriety of taking any specimens, consult appropriate fish and game wardens. These officials can be helpful in locating a suitable specimen, assuming, of course, that the request is valid, legitimate, and cannot otherwise reasonably be satisfied.

PRESERVING SMALL SPECIMENS

Small mammal specimens, fish, reptiles, frogs, and other amphibia can be preserved whole in a solution of one part (40% commercial strength) formaldehyde (formalin) to nine parts water, or in alcohol (about 75% strength). Keep them in labeled glass jars for easy observation.

BIRD SKINS

Prepare bird skins by skinning them in the same manner as described for mounting. Skins may be preserved by applying borax to their inner surfaces and bones, but for long-term protection they should be poisoned. Tamp wads of cotton in eye sockets, skull cavity and throat, then tie the mandibles shut with sewing thread passed through both nostrils and knotted underneath. Pad and tie the wing bones together in natural position within the skin. Replace the leg muscles by wrapping the retained leg bones with cotton batting. Then spin the correct amount on a thin, smooth, pointed stick, forming a somewhat flattened artificial body form of appropriate length. Thrust the thin neck portion well up into the head and insert it in the study skin. Then withdraw the stick and close the skin incision with a few stitches of thread, being sure to pick loose any feathers caught in the stitches. Preen the feathers almost constantly while doing the work. Tie the feet together in a crossed position and position the wings neatly in place before wrapping the entire specimen lightly with gauze bandage material. Turn the heads of crested birds to one side. (See illustrations.) Some collectors

Mammal and bird study skins *(left)* pinned and wrapped for drying; *(right)*, study skins, dried and tagged.

prefer a stiffener in bird skins. Ordinary pipe cleaners or prepared wooden rods may be used for this purpose. These can be wrapped in along with the stick and left in place when the stick is withdrawn.

When the skin is completely dry, remove the wrapping, repreen the feathers and attach a tag to the legs containing all pertinent data. Listed should be the species, date of capture, sex, locality where taken, measurements, collector's name and other details.

When sex cannot be determined externally, the gonads must be examined by opening the abdomen. Place the carcass on its back and inspect the sex organs that lie along the backbone adjacent to the kidneys as shown in sketch.

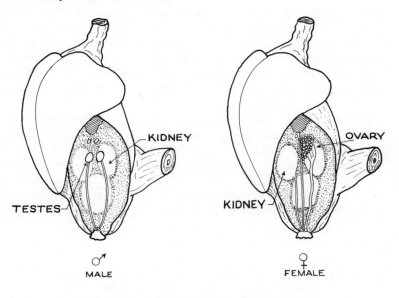

Determining sex of bird by examination of gonads.

SMALL MAMMAL SKINS

Small mammal study skins are handled in a similar manner, except that the skull is kept outside for scientific examination. Be sure to clean the skull carefully and keep it with the skin. Poison the skin and bones. Wrap the retained leg bones with cotton or tow to their natural size. The legs of the larger specimens should be wired for support. Spin a thin covering of cotton onto a sharpened wire, moisten this with borax water and insert it into the tail skin as was done in mounting. The tail wire is made long enough to extend through the body into the head skin where it is bent back in a tight loop. Now fill the head and body skin loosely with cotton and sew up the incisions. Apply a few stitches to the mouth opening.

Shape the specimen as naturally as possible with the fingers, the specimen resting in straight, belly-down position, its legs extended to front and back. Brush its fur and pin the skin to a flat surface until dry. Attach the skull and identifying tag detailing complete data, as for bird skins. The label should be securely tied just above the heel of the right hind foot as shown in sketches.

HANDLING TRAPPED OR OTHER LIVE SPECIMENS

Live birds may be dispatched humanely by holding their heads down in a cone of paper containing a wad of cotton saturated with chloroform. Mammals, too, can be handled in this manner, using an enclosed box or plastic bag instead of a paper cone. Animals like coon or fox are quickly dispatched by drowning, or by stepping heavily on the chest just back of the foreleg, a method occasionally resorted to by trappers.

Use wads of cotton to plug up immediately all openings in the bodies of specimens, especially those that have been shot. Mouth, nostrils, ears, anus and shot holes should be securely plugged to avoid possible stains to fur or feathers. It should be remembered that blood is much more easily removed when fresh. Use cold water to remove blood stains. Clean and wrap specimens carefully in paper before putting them into the game pocket.

RECORD KEEPING

A field catalog or notebook listing the specimens collected should be kept with each specimen being numbered. This number should appear in the catalog, on the label attached to the skin, and on the skull label. Pertinent information not appearing on the label can be noted in this book. It is important that all data be authentic! Unlabeled or inaccurately-marked specimens have little scientific value.

CHAPTER 9

An Introduction
To Tanning

MODERN COMMERCIAL facilities have practically eliminated the necessity for anyone to do his own tanning, especially since so many good tanneries in all parts of the country do such a superior job at a very reasonable cost. Fast mailing services make it possible to send raw skins and hides anywhere in the country in perfect condition for expert processing, where the job is done at a price anyone can afford to pay.

IMPORTANCE OF CORRECT PROCESSING

The value of any skin depends much upon the manner in which it was handled in the raw state. Correct skinning and curing of all skins and hides is imperative if true value of the finished product is to be obtained.

Most fur skins are used for the manufacture of luxury items and as such are valued largely on their beauty and pleasing appearance. Furthermore, high-quality furs are very valuable and if handled with care, find a ready market. Very probably, one should sell such skins to the trade, where a good price can usually be demanded, rather than attempt to profit from them through amateur processing.

The amateur should not attempt to tan valuable fur skins or hides for home manufacture of coats, jackets, gloves, etc., since the risk involved in producing quality products as measured by usual standards of quality and appearance is too great. A great deal of experience is required in the manufacture of leather and fur garments and it is doubtful if more than a few could qualify to produce anything in this line that would be acceptable by the ordinary standards.

Tanning, moreover, is only the first step in production of the finished article. Skins must then be blended, dyed, sewn together and tailored. These operations are all highly specialized skills and would be quite impossible for the amateur to accomplish, therefore, such work is best left to those who are experienced and equipped for it.

However, from the standpoint of usefulness and serviceability, inexperienced persons desiring to do their own tanning and simple leather work much as did their forefathers, the pioneers, may achieve a fair degree of success through the tanning formulas and processes here described. Many people take justifiable pride in doing their own tanning work, and do derive much pleasure from experimenting with various formulas and methods.

Successful tanning depends on a great diversity of factors such as the time element, amount and strength of formula ingredients, condition and character of raw materials and many others, each dependent upon the other. First-class results in the tanning and leathering of animal skins and hides by the amateur are only possible by more than a little manual effort. Work with the fleshing knife and at softening the hardened skin fibers over a fleshing beam requires a good deal of conscientious effort. Incomplete scraping of the flesh side hinders penetration of the tanning fluids. Pliability and softness can only be accomplished by much "working" or "tousing" as it is called.

It will be noted that this book describes a number of processes ranging from the simple preparation of rawhide and dry curing of skins to the more involved methods of leather making. Some of the procedures are similar in character. Each has proved successful under certain circumstances. While it may seem that many of the procedures are repeated unnecessarily, this was done only to stress their importance. It may be well to experiment with several of the processes, or vary them to some degree, since one or another should achieve the individual results desired.

Coverage of tanning here is intended chiefly as a guide for those who wish to experiment in this work in an amateur way. It is suggested that the amateur not use any valuable fur skin or hide in his initial attempts at tanning, as considerable experience in the work is required before good results are obtained.

Also, it should be remembered that some animal diseases, such as tuberculosis, tetanus, rabies, anthrax, etc. can be transmitted to humans through their handling of infected animals. Any animals found dead, acting peculiarly, or in a sick and weakened condition should be suspected of being a potential carrier and should be destroyed immediately.

HISTORICAL NOTES

The history of tanning goes back to the very origin of civilization itself. In Genesis III, 21, is written "Unto Adam and also His Wife did the Lord God make coats of skins and clothed Them." The early Egyptians and Hebrews have written much about the uses and methods practiced by the ancients. Direct records date back to the time of the building of the pyramids, nearly 5000 years ago.

An early recipe recorded by the Arabians is as follows: "The skins are first put into flour and salt for three days, and are cleaned of all the fats and impurities on the inside. The stalks of the chulga plant, being pounded between large stones, are then put into water, applied to the inner side of the skin for one day; and the hair having fallen off, the skin is left for two or three days and the process is completed."

The early Greeks and Romans contributed much to the science of skin tanning, some still the basis for modern methods. The word "tan" was coined by the Romans who used leather and tanned skins as a basis for money, hence the word "pecuniary," a derivative of the Latin word for "hide."

The first settlers in America found that the Indians were well versed in the process of skin tanning. The Crow and Navajo tribes were especially adept in skin dressing, and taught the pioneers the process of soft "Buckskin Tan" which even today has scarcely been improved upon. The Indians made fleshing knives of certain leg bones of the animals they killed and used a sloping, smooth tree trunk as a fleshing beam. The hair was either removed by allowing the skins to decompose or by applying a solution of lye made from their campfire ashes, after which the skins were rubbed with a composite mixture of brains and liver. The hides were then pegged to the ground and allowed to dry, after which they were manipulated by hand until soft. The skins were then smoked over a small fire and became thoroughly soft and water-repellent.

Up to the latter part of the Eighteenth Century, the tanning agent used by most civilized tanners was oak bark. The hides were soaked for six months or more in vats of water with layers of crushed oak bark between them. About 1800, Sir Humphrey Davy discovered that various other plants contained tannin such as sumac, hemlock, mimosa, and chestnut barks as well as the fruits of the divi-divi, volania and others. About one hundred years later an American chemist named Schultz discovered the processes of chrome tanning which were later perfected by a native of Philadelphia, Robert Foerderer. His experimentation with the chromium salts together with the soap-oil "fat-liquor" process, which in reality is only a modification of the old Indian method of softening a deerskin by dressing it with fat, gave the industry its greatest modern development.

Mineral tanning with metallic salts has largely replaced vegetable tanning for all but a small percentage of heavy leathers. Modern processes and machinery have speeded-up production and greatly affected the industry, consolidating it into a relatively small number of large concerns, each specializing in a particular type of leather production. Most of the tanneries are located in large cities and seaports, where they are closer to world markets and transportation facilities.

The old tanneries, once familiar landmarks in almost every sizable community, have almost disappeared from the American scene. With them vanish the knowledge and skills of the oldtime tanners.

The development of synthetic materials, plastics and other substitutes has narrowed the uses of leather. It is an interesting fact that 4/5 of the leather produced in America is now used in the manufacture of shoes, most of the balance going into gloves, jackets, novelties and industrial requirements.

NEEDED TOOLS AND EQUIPMENT

The tools required for tanning are few. After familiarizing himself with the text covering the various skinning, dehairing, fleshing and beaming operations, the amateur tanner should be able to select the tools necessary for his particular job.

Substitute tools can be fashioned as suggested, but it is better to purchase the correct tools from a reliable dealer.

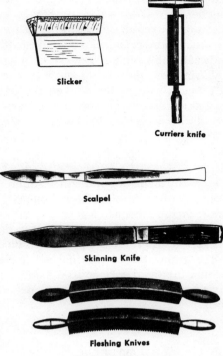

Tools used in tanning.

Knives

For general skinning, a slim, pointed blade is most desirable. Broad-bladed butcher knives are best for fleshing and trimming work. With any knives, one should always have a knife steel or whetstone for sharpening.

Currier's Knife

This is a specialized, double-edged tool designed for fleshing and scraping larger hides over the fleshing beam. Both sides are kept sharp for shaving and thinning the skins. Provision is made to permit blade changing. The handles are positioned so as to allow complete control of the blade angle. This tool is a convenience, but not an essential.

Fleshing Knife

This is a two-handled draw blade available in several styles. (See illustration.) Hand fleshing knives can be made from machine hacksaw blades (ordinary hacksaw-blades are too light), carpenter's draw knives, butcher knives, or even scythe blades, but it is recommended that the amateur purchase these special tools as they are inexpensive and much more practical than makeshift substitutes.

A person handy with tools can fashion a vertical-type, fixed blade for small-skin shaving. An old saw blade with one edge curved and sharpened, is held vertically by brackets bolted to bench and wall. The operator sits directly in back of the vertical blade and shaves the skin by drawing it across the sharpened edge (see "Pulling Bench").

Slicker

This simple tool is easily made out of a 5-inch square piece of steel or brass about one-eighth inch thick. Round one edge slightly and fit a wooden handle on the other as suggested in the tool illustrated.

A hardwood block about 6-inches square, tapered wedge-like to a smooth thin edge will also serve the purpose. The slicker is used to smooth out the grain side of the finished leather, eliminate wrinkles and scours the oiled surface to a smooth finish. It also serves in removing excessive moisture from the unfinished furs and leather goods.

Fleshing Beams

Fleshing beams serve several major purposes for the tanner. They provide a smooth, rounded surface on which skins may be fleshed under complete control of the operator and their sharply-rounded tips break down the hardened skin fibers when skins are "worked" over the edge. The size and shape of the beams are determined largely by the characteristics of the skins to be worked on.

For small fur skins up to the size of fox, raccoon, etc., a small bench beam, shaped as shown in sketch, should be sufficient. Use a hardwood plank about 18" in length. 1½" thick and 4" wide. This type may be clamped or bolted to a bench where it can be swung out of the way when not in use. Bolt larger beams for heavier skins and hides to a framework of 2 x 4 stock in a sloping position, height and length being determined by the individual build of the user. Make this beam about 6' in length, 10" wide and 2" thick. The tip should be just above the waist of the currier when he is in a standing position. The beams should be whittled from maple or other hardwood and should be tapered wedge-like to a rounded point not more than one-eighth inch in thickness. Some beams are designed so as to be adjustable for height and length.

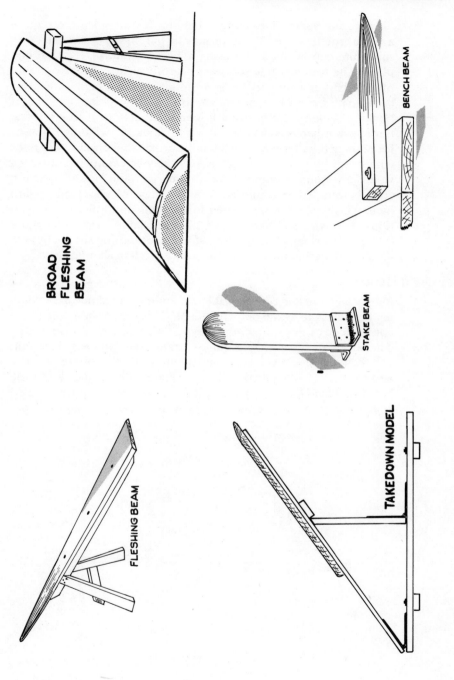

BROAD
FLESHING
BEAM

BENCH BEAM

STAKE BEAM

FLESHING BEAM

TAKEDOWN MODEL

Typical fleshing beams.

Large and heavy hides to be tanned for leather purposes require a much broader beam. This type may be purchased or a satisfactory substitute can be made by using a broad plank or log slab and planing its rounded side to a very smooth surface. Place one end of this slab on a support so as to be about waist high. Since this broad-type beam does not taper to a narrow point, an upright "staking" beam must be used in conjunction with it (see illustration). Use a hardwood board about 1" thick, 6" wide and 3' long for making this. The stake beam must be solidly bolted to the floor with heavy blocks or angle irons, or it may be clamped in a strong vice. Some tanners simply drive it solidly into the ground. The top edge is rounded and tapered to a thickness of about one-eighth inch. This type of beam is used for fleshing and dehairing operations. The skins and hides may be made pliable after tanning and drying by working them back and forth over the edge of the beams, much in the same manner as one uses a cloth in shining shoes.

Pulling Bench

People who tan and finish a sizable number of fur skins use a "pulling bench" such as the one illustrated. This simple bench is easily constructed of scrap lumber, a piece of pipe and an old saw blade. The exact dimensions are unimportant, the bench being built chiefly for the convenience and comfort of the operator. The vertical post can be mortised into a square hole on one end of the bench and held firmly in place by driven wedges. The post is drilled near the top to receive one end of a threaded steel pipe or rod.

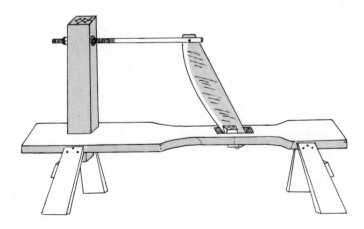

Pulling bench.

Large nuts and washers provide adjustment for the blade angle. The other end of the pipe is slotted to admit the upper end of the blade which is fastened with a single bolt. The lower end of the blade extends downward through a wide slot in the bench top to a point between the currier's knees, and it is held rigidly in place with wooden wedges driven on each side.

The blade, sharpened edge away from the user, should be about two feet in length, six-inches wide at the belly, and narrowed to about four inches at the ends. Slant the blade about as shown. The bench may be notched on its sides to allow comfortable clearance for the operator's knees. This upright knife assembly can also be mounted on a workbench and fastened to the wall.

With practice, the amateur will learn the art of drawing the tanned skins across the sharp edge without cutting into them. This is done with the dry skins both before and after oiling them. Nothing else will do as good a job of breaking down the tough skin fibers to admit oil or smooth the oiled skins to as soft and velvety a finish.

SKINNING AND HANDLING

A few words on skinning might be of interest to tanners, farmers, trappers and sportsmen in general. Animal skins destined for the fur trade, such as weasel, mink, marten, fisher, fox, oppossum, muskrat, civet, skunk, and wild cat should be skinned "cased" (see illustration). The hind legs are ripped from heel to heel via

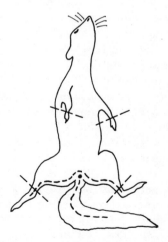

Cuts for case-skinning furbearers.

the root of the tail. Cut around the vent and skin the back legs. The tail should be skinned down a short distance, then stripped, either with the fingers or by holding the tail bone between two round sticks. When the tail bone is removed, the tail skin should be ripped its full length on the under side.

Hang the animal up by the hind legs and peel the skin from the body using the skinning knife where necessary. Sweeping strokes slightly towards the skin help remove close clinging fat and tissue. Try to remove as much of this as possible without damaging the skin. This is especially important in the case of beaver, raccoon

Skinning gambrel.

and other very fatty animals. The front legs are then slit from foot to shoulder on the under side and skinned. Peel the skin over the head, cutting closely to skull at eyes, ears and lips. The pelts are then inverted on drying frames or boards, flesh-side out until dry, with the exception of fox and lynx which are reversed while still flexible, and allowed to dry with fur side out. Dry, cased skins are slit up the belly before tanning. (See accompanying sketches.)

Animals like raccoon, bear, beaver, badger, wolf, mountain lion, wolverine and coyote are generally skinned flat or "open." (See illustration.) They are stretched by lacing with strong cord or rawhide to a frame or hoop. Beavers are properly stretched round and raccoons nearly square. All other skins are stretched to their natural shape.

The losses to trappers from lack of knowledge in skinning, curing and handling furs for market or home tanning are considerable. Many skins reach the dealer in poor condition and are graded "small" or inferior. Some become tainted when improperly dried and packed or are stretched out of shape on ill-fitting boards. Many

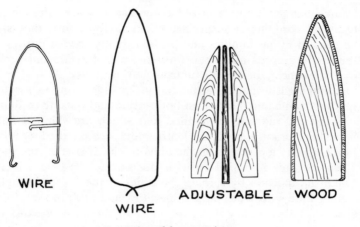

WIRE   WIRE   ADJUSTABLE   WOOD

Examples of fur-stretchers.

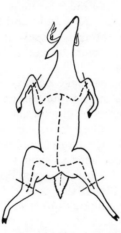

Cuts for flat-hide take-off. Horned heads are skinned through an incision made in the back of neck.

fine pelts are ruined by exposing them to the hot sun which soon burns the life out of them. Others are hung up by the nose, the weight of the board drawing the fore part of the skins out of shape and rendering them worthless. Trappers who handle their furs in this manner can never expect to receive full value for their furs.

All furs should be dried in a cool, airy place, never near a fire or other source of intense heat, as this causes them to dry out too rapidly and become brittle and unfit for tanning. Fur and hair skins should not be allowed to remain for a long time on the drying frames, but should be removed as soon as they are sufficiently dry to avoid wrinkling or shrinking. When shipping furs, they should be packed in flat bundles and packed tightly. Avoid cutting with the binding cords and keep the ones with fur-side out separate from the others, to keep the fur from getting greasy. A good method for shipping is to pack the furs in burlap bags, using tags with the sender's name and address on both inside and outside of bundles.

Skins and hides to be tanned and which are skinned flat for buckskin, fur rugs, rawhide, leather, etc. are handled as follows: Make a belly incision from the anus to the corner of the mouth. Then each leg is ripped on its under side to the center incision. Make the incision by forcing the point of the knife through the skin or hide with sharp edge uppermost and push in the direction desired. This avoids cutting the hair. In the case of skins to be tanned for rugs, chokers and for taxidermy purposes, the feet are left on, in which case they must be incised on their under sides and all flesh removed from foot and hand bones. When skins are to be leathered for mounting purposes, leg bones and skulls are usually retained and care must be taken to avoid cuts and tears, especially about the face.

It is important to cut cartilage of ears, eyes and nostrils close to the skull. On animals to be mounted, only the lineal incision is made, that is, the lengthwise cut on the underside of the body (from point of chest between front legs to the vent). For more information on this subject, refer to Chapter 5. Ears must be skinned and inverted over the cartilage to their tips, lips must be split to their edges and the entire face at eyes, lips and nostrils must be shaved as thin as possible. If this is not done properly, hair-slip will result, because the tanning liquids will be unable to penetrate the tissue at these points.

Remember, unsalted hides and skins, thrown carelessly together, generate heat and are quickly rendered worthless for any purpose in a short time. Deer and similar hides can be dried when they are spread individually in a shady, well-aired spot, but they must never be exposed to the direct rays of the sun as they burn easily. Trim off tail, head, shanks and ragged edges of hides to be tanned for leather purposes.

SALTING AND CURING

Pelts are not usually salted, but are carefully stretched and dried to their proper shape. Use an old kitchen spoon to scrape fatty skins on the stretcher board to insure drying. Some salt may be necessary on the fleshy side of ears and nose parts only, on very fatty skins.

Larger hair skins, such as deer, bear, etc., are treated in an entirely different manner. The following description of handling a fresh "green" deerskin for later home tanning or shipment to commercial establishment, applies equally well to other similar-type skins.

After having completely removed all vestiges of flesh, fat, dried blood, dirt, etc., as well as tail bone and dew claws, trim off the ragged edges of shanks and head. Spread the skin out flat, flesh-side up and pour a liberal pile of clean fresh salt in the center. Use at least one pound of salt per pound of hide. Salt is cheap, so use plenty! Rub it in well, using the hands, working to the edges. Every part must be well covered. Then fold the skin flesh side to flesh side, roll it up and place it on a sloping surface to drain. In a day or so, unroll it, shake out the old salt and repeat the operation, resalting well. The hide should be drained again, if necessary. Then spread it out flat and allow it to dry in a cool, airy place.

Sometimes a moose, or elk hide is so thick at the neck that it may be necessary to make criss-cross cuts about 1/2-inch apart and halfway through the skin on the flesh side of the heavy area. Rub salt into the cuts to finish curing, since salt is only effective to a depth of about 3/8 inch.

To insure return of the same hide from the tanner, it is advisable to mark the raw skin by punching a series of holes in a recognizable design, preferably along the edge on the thicker butt or neck portion of the hide.

If the skin is to be shipped to a commercial tanner for leathering, salt it as described, fold the skin once, flesh-to-flesh, and roll it up while still damp and flexible; wrap it well in several layers of burlap and bind it with twine. The bundle can be mailed in a suitable carton box together with a letter of instruction to the tanner. Be sure to include the name and address of the sender, both inside and outside of the container.

SOAKING, DEGREASING AND FLESHING

Dried skins to be tanned with fur or hair on are soaked and softened most effectively in water to which soda or borax has been added. Use about 1 oz. to the gallon of warm water. The water used in soaking should always be fresh and clean. Any soaking that is done should be accomplished as quickly as possible to avoid decomposition and loosening of the hair. In the case of greasy skins it may be necessary to soak them in borax water first and then wash them in a warm sal-soda solution.

It is important that all water used in the tanning and soaking processes be of the soft type. If the tanner operates in one of the many regions of the country having hard-water conditions, he must resort to the use of rain or cistern water sources or install special water-softening equipment. The mineral and chemical content of hard water usually has an adverse effect on the tanning processes.

Some tanners use the older-type washing machines for soaking and rinsing skins. The gentle paddling action does a fast and thorough job.

Beaver and similar-type water animals are naturally well fortified with layers of fat, and in skinning these animals it is almost imperative to skin in a manner so as to allow most of the fat to remain on the carcass. Such skins must be degreased before tanning.

To do this, soak small skins in naphtha solution or pure white gasoline. On larger skins, sprinkle the naphtha on both hair and flesh sides and rub with warm sawdust. Work the sawdust well into the hair or fur to absorb the naphtha and grease. Do this work in a well-aired place away from any flame, or outdoors, if possible. The skins are now hung out in the air and the sawdust beaten out with a small pliable switch, after which the skins are washed and rinsed in clear water.

Furs can also be cleaned by rubbing them with a wad of cloth soaked in white gasoline or benzene, after which they are washed and rinsed in warm water and "fluffed" with powdered borax or plaster of paris. Beat out the powder when dry with a flexible stick. An electric hair dryer or the blower attachment to a vacuum cleaner does a good job of drying and fluffing fur pelts.

All flesh, fat, bits of sinew, etc., must be removed from hides and skins before tanning. This includes the thin, tight layer of tissue found on all skins. The tissue must be broken up and entirely removed, which is best accomplished by alternately working and

This effective fleshing tool can be hand-forged by a blacksmith. It is important that it hold a sharp cutting-edge at just the right angle.

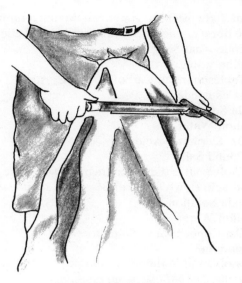

Fleshing a large hide.

soaking. A sharp, finely-toothed metal edge can be used to scratch up the tenacious membrane after which it can be scraped off entirely.

At the same time, much oil and grease are worked out of the skin, especially when the work is done on a suitable fleshing beam. In doing this scraping and fleshing, care must be taken to avoid scoring or cutting the skins, which easily happens when using sharp tools on the fleshing beam. Also, do not injure the true skin or expose the hair roots in this work.

FORMULAS FOR TANNING SOLUTIONS

Caution should be observed when mixing and handling the chemicals since many of them are caustic and poisonous. Always mix and store tanning liquors in glass or earthenware containers. *Never in metal!*

Most of the larger towns and cities have firms specializing in the manufacture or production and handling of chemicals. Druggists generally have connections with chemical houses and can supply ordinary requirements. Many taxidermy supply houses advertising in sporting magazines make up various tanning formulas and offer a good selection of tanning tools and supplies.

Names and addresses of dealers and manufacturers of tanning tools and chemicals may be obtained from MacRae's Bluebook or the Thomas' Register of American Manufacturers, available in any public library, where they are listed alphabetically by subject.

1. Carbolic Acid Solution
   1 Gallon water
   1½ Tablespoonsful Carbolic Acid Crystals
   Mix as needed.
2. Pickle Solution
   1 Gallon Water
   1½ Oz. Sulphuric Acid
   1/2 Pound Salt
   Dissolve salt in the water, then add sulphuric acid. Caution... add acid slowly. *Do not use metal container!*
2A. Pickle Solution
   1 Gallon Water
   1/2 Oz. Concentrated Sulphuric Acid
   1 Pound Salt
   Dissolve salt in the water and pour in the acid slowly while stirring. *Do not use metal container!*

3. Neats-Foot Oil Solution
   1 Part Water
   1 Part Sulphonated Neats-Foot Oil
   Should be warm when prepared and used.
4. Alum-Carbolic Acid Solution
   1 Gallon Water
   1/2 Pound Salt
   1/4 Pound Alum
   1/2 Oz. Carbolic Acid Crystals
   Use warm water and stir thoroughly, each time used.
5. Dehairing Solution
   1/4 Cup Lye
   10 Gallons Water
   Use caution in mixing. *Wear rubber gloves when handling.* (*Note*—See also method of dehairing hides with lime water solution as described under Chrome Tanning— Heavy and Medium-Weight Leather.)
6. Oxalic Acid Solution
   1 Gallon Soft Water
   1 Pint Measure Salt
   2 Oz. Oxalic Acid
   Poisonous! Heat part of water and dissolve in it Salt and Acid crystals.

7. Salt-Alum Solution
   1½ Gallons Water
   1 Pound Ammonia Alum (or Potash Alum)
   4 Oz. Washing Soda (Crystallized Sodium Carbonate)
   8 Oz. Salt
   (Prepare solution as described in text.)

7A. Salt-Alum Solution
   10 Gallons Water
   5 Pounds Salt
   2 Pounds Alum
   Dissolve ingredients in warm water. Allow it to cool before using.

8. Dry Curing Formula
   2 Parts Alum
   1/2 Part Saltpetre
   1 Quart Measure Salt
   Mix chemicals thoroughly.

9. Chemicals Required for Chrome-Tanning Heavy Hides (over 30 lbs.) For Leather.
   12 Pounds Chrome Alum (Chromium Potassium Sulphate Crystals)
   3½ Pounds Soda Crystals (Crystallized Sodium Carbonate)
   6 Pounds Salt (Sodium Chloride)
   Use only the clear, glass-like crystals of Alum and Soda. (See text for preparation of solution).

10. Combination Solution
   1 Pound Aluminum Sulphate
   1 Pound Salt
   3 Oz. Gambier (or Terra Japonica)
   Dissolve Aluminum Sulphate and Salt together in a little water. Dissolve Gambier in a little boiling water, then mix both solutions and make up to 2 gallons with water.

CHAPTER 10

Developing
Tanning Skills

FOR THOSE WHO WISH
to try their hand at tanning, there follows a description of the most
suitable processes together with procedures for making rawhide,
buckskin, etc., and handling furs.

TANNING PROCESSES FOR LEATHER AND FURS

Unless otherwise indicated, all references to tanning soaks or
solutions are to those given in the preceding chapter.

Oxalic Acid and Small Skins

A cheap and simple way to tan squirrel, skunk, rabbit and similar
small animal pelts, is to use formula #6. Use the solution carefully, as it is poisonous. Soak small skins in this solution, stirring
occasionally, for about 24 hours. Medium-sized skins require about
three days. When the hides are tanned, remove them from the acid
solution and soak overnight in a mixture of 1/2-cup sal soda and
five gallons water. Rinse again in clear water to remove all traces
of salt. The rinsing will prevent any "sweating" later. Hang the
wet skins up until almost dry, then scrape and work them with
knife and beam as described for buckskin. If the small hides dry

hard while working, dampen them again and continue working them until they become soft and supple.

Salt, Alum and Fur Skins

This is an old and widely-used method of fur skin tanning used by many taxidermists, usually resulting in pelts with good stretch and flexibility as well as greater durability. If not done correctly, however, first attempts may come out stiff and hard, in which case re-tanning and repeated beaming may be necessary.

Make up the tanning solution #7 as follows: Dissolve the pound of alum in one gallon of water; dissolve the soda and salt in the half-gallon of water; then pour the soda-salt solution very slowly into the alum solution while stirring vigorously. Clean, flesh and soak the skin as previously described. Immerse the skin in the solution from 2 to 5 days according to its thickness. Move the pelt about in the solution occasionally, squeezing and turning it periodically. At the end of this soaking, remove the skin and rinse it in a borax-water solution, using about one ounce borax to the gallon. Then rinse it completely in clear water, using several changes. Squeeze out most of the water, then stretch it out smoothly, flesh-side up, and tack it to a flat surface. Apply a thin soap paste to the skin with a cloth or sponge and allow this to be absorbed. After this has gone in, apply a thin coating of Neats-foot or similar oil. Allow the skin to remain until almost dry, then while still slightly damp, work it over the beam, rewetting and towsing repeatedly as necessary. Finally, clean in naphtha or gasoline and warm sawdust, beat, comb and brush.

A slight variation of this method of tanning with salt and alum makes use of formula #7A. Dehaired hides of any size may be handled by using this solution. Allow the tanning solution to cool before immersing the hides. Stir several times a day while soaking. Large hides take approximately six days, the smaller ones less time. When tanning is complete, rinse the skins thoroughly to remove all traces of the salt brine. (It is often possible to tell if tanning is complete by cutting a sliver of the skin from the edge and examining it. If the cut shows the same light color throughout its thickness, the tanning is finished.)

Hang the skin up until almost dry, then work the flesh side vigorously over the fleshing beam. When the skin is dry, dampen it and roll it up until pliable, then repeat the work. This must be done several times until the skins remain as flexible as cloth.

The final step is to dampen it slightly and rub vigorously with a little warmed Neats-foot oil. Then work it over the beam once more. The hide should remain supple, even after becoming wet in use.

Alum-Carbolic Acid Soak

This method is similar to Salt and Alum tanning. Large dried skins must be soaked in carbolic acid solution #1 to relax them. Small skins may be relaxed with the solution applied to their flesh side and by sprinkling them with sawdust. When the skins are pliable, immerse them in solution formula #4. After about six days, they may be removed and drained. When thoroughly drained, apply a coat of sulfunated oil (formula#3) and allow the skins to dry. Then relax the skins again with carbolic acid-water solution and work them over the fleshing beam. After this, shave the skins again, gradually working them dry over the beam until pliable. Skins will be very soft and white in color when properly finished.

Alum Paste and Small Fur Skins

Small furred animal skins such as rabbit, coon, skunk, squirrel, etc., which are to be used as glove or garment linings, rugs, etc., and which will receive wear can be tanned successfully with an alum paste prepared as follows:.

Dissolve the dry ingredients listed in formula #7 in a little water and stir in enough flour to make a smooth, thick paste. Apply the paste to the flesh side of the skins and let the paste remain on for several days, then scrape it off and apply a fresh coating. Do this three or four times, then rinse and work as described in the salt and alum tanning method. This method tends to shrink furred skins, thus aiding in hair retention, a desirable characteristic in utility furs.

Pickle-Neat's-Foot Oil Tanning

Assuming that skins are thoroughly fleshed, free of grease and soaked, proceed as follows:

Lay out the skin flesh-side up, and apply a coating of pickling solution, formula #2. A sprinkling of sawdust may be placed on the skin, which prevents the liquid from running off. Allow the skin to remain in this position overnight or longer, but do not allow it to dry out. Then remove the sawdust and coat the skin thoroughly with Neats-foot oil, formula #3.

Hang the skins up until thoroughly dry. Now dampen them with a solution of carbolic acid (formula #1) and roll them up, flesh-side in. The following day the skin is ready to be "toused" or worked over the fleshing beam. Again, the more effort put on the beam work, the better the result. If the skins or hides are heavy, or not thoroughly soft and pliable, dampen again with carbolic acid water, roll them up overnight and repeat the beam work. Stretch the skin out flat and apply a coat of warm, soapy water, roll it up and allow it to remain overnight, after which it is given another light coat of the oil.

Salt-Acid Tanning

This old process of tanning, is properly called tawing, and it is simply a variation of the Pickle-Neats-foot oil tanning as described before. Use caution in handling the sulphuric acid, making sure that the fumes given off are not inhaled, and avoiding exposure of the skin or clothes to the strong acid.

Using the ingredients of solution #2A, dissolve the salt in the water and carefully pour the acid in very slowly while stirring. Use a glass, wooden container or earthenware crock to mix and store this solution, *never metal*. When the mixture has cooled, it is ready for use. Put the clean, softened skin in the solution, covering it completely. It may be necessary to weight down the skin or skins, using a good-sized boulder or rock.

Any number of hides or skins may be soaked in this manner. Allow the skin to remain in the solution from one to three days, depending on its weight or thickness. Stir and move it about periodically, making sure that no air is trapped beneath the skins. After tawing, rinse the skin in clear water and squeeze it out without wringing. Then work the skin for 10 to 15 minutes in a borax solution (1 oz. borax to the gallon), after which rinse it again in clear water and squeeze it out. Slick out the skin on the flesh side and allow it to remain until almost dry. Then, while still damp, work the skin over a beam or stake as described before, until soft and supple.

Scraping the skin with a dull knife or sandpapering it will be of help in smoothing the flesh side and aids materially in the softening process. Skins tanned by this method tend to become damp and clammy in damp weather and, if wetted repeatedly, lose their tanned effect, as is the case with similar acid tans.

Combination Mineral and Vegetable Oil
Tanning For Fur Skins

This process is very popular and successful when properly done, resulting in lasting soft and pliable tannage of fur skins. After the skin has been thoroughly cleaned and softened, it must be fleshed and scraped with a dull steel blade such as an old knife having notches or teeth filed in the blade. The layer of thin tissue found on all skins must be broken up and entirely removed. This membrane must be worked on with the tool, scratching it up and scraping it off after repeated soakings. All grease must be worked out of the skin until it is uniformly soft and pliable. In this fleshing and scraping operation, avoid damage to the hair roots and true underskin, especially on thin, delicate skins.

After the skin is free of grease, soak it for several hours in a warm borax or soap solution, using about an ounce to the gallon of water. This treatment promotes softening and cleansing of the epidermis. Now squeeze out the excess liquid and hang the skin up for a few minutes. It may be well to dry out the fur-side somewhat before tanning, by working in a little warm sawdust. If in drying out the fur-side of the pelt, the flesh-side becomes too dry, dampen it with a wet cloth or sponge before applying the tanning paste.

Using solution #10, prepare a moderately-thin paste by adding ordinary flour to a quantity of the liquid. (In some cases, such as skins with little or no natural grease or of a tough texture, a little glycerin or olive oil may be added.) Tack the skin out flat and smooth, flesh-side up. Apply two or three coatings of the prepared paste at daily intervals, depending on the thickness of the skin. Only fairly heavy skins require three coatings. Make each coating about one-eighth inch thick and apply at intervals of a day. Keep the skin covered with cloths or paper between applications. Scrape off most of the old paste coating before applying a new one.

After the last coating has been applied, allow the skin to dry. When almost dry, wash off the flour paste and rinse for several minutes in borax water, using an ounce of borax to the gallon. Then rinse in clear water alone. Squeeze out most of the water and place the skin on the fleshing beam and slick it out well, removing most of the moisture. Then tack it out on a smooth surface, flesh-side up, and apply a thin coating of Neats-foot or similar-type oil. If the skin was of the greasy type, it may not be necessary to apply any oil.

When nearly dry, but still damp, work the skin over a fleshing beam or stake, as described under "Beams." The important thing here is to start the beaming at the proper time. The skin must not be too wet or too dry. Experience will indicate the best point at which to begin this work. One indication of this is the appearance of light spots on the skin when folding or stretching it. If the skin becomes too dry during the staking process, it must be dampened evenly and worked again while drying. This must be repeated several times, if necessary.

Finish the fur skin by working it in a quantity of warm sawdust or powdered borax, then shaking it out, switching, combing and brushing it. Sandpapering the flesh side of the skin will help smooth it out and this also helps in thinning the heavier portions and making the whole skin more pliable and flexible.

MAKING RAWHIDE

Rawhide has many practical uses and it is easy to prepare. It is extremely strong and durable and provides excellent strap and lacing materials for boots, snowshoes, etc. Wet rawhide will mold easily over wooden forms and dries to any desired shape. Rod and gun cases, saddle-bags and the like are easily made in this way.

Deer and elk hides, as well as wolf, coyote and even woodchuck are used to make the best rawhide. Soak the skin for a time in clear water to soften it, then put it in the dehairing solution #5, overnight. Using rubber gloves, stir the hide occasionally during the next day and test it several times for loose hair. When the hair slips off easily, scrape it away with the thin skin in which it grows, using a dull knife or other blade. Use the fleshing beam for this work. The flesh side must now be thoroughly scraped to remove fat, gristle, bits of flesh, etc., after which it is stretched tightly, either by tacking it solidly on a flat wooden surface in the case of small skins, or laced inside a frame of poles. Punch holes at two-inch intervals around the edge and lace the skin tightly with strong cord or rawhide. Make sure the hide is stretched to the maximum. It can then be set aside until dry when it is released and stored in a cool dry place until used. Soak the rawhide in water until soft, before cutting or molding. Boot laces and similar rawhide products are oiled well with Neats-foot oil to make and keep them in a flexible condition.

DRY-CURING SKINS

This is about the simplest way to preserve a small skin for a flat trophy. It does not result in as soft and pliable a skin as other methods, yet it may be satisfactory for some work.

Wash the pelt well in warm soapy water, then dry it with sawdust, powdered borax or plaster of paris. When the fur is dry, beat

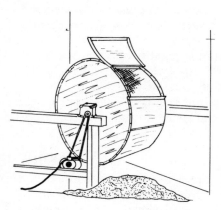

Homemade drumming apparatus.

Fleshing machine.

the drying agent out with a small switch and comb the fur well. Lay the skin down flesh-side up and apply the combination dry-curing formula #8. Rub the chemicals in well, making sure all portions are covered. Fold the skin once, flesh-to-flesh, then roll it up and put it away overnight. Examine the skin the next day and if there are no traces of fresh blood or pink spots, the preparation is working satisfactorily. Give the skin another rub-down with the chemicals and roll it up and store it for another 24 hours. After the skin has been cured, it must be made as soft and flexible as possible by scraping it with a dull knife and working it over a fleshing beam. (Indians and pioneers hammered a long stake in the ground and worked the skins over this.) With plenty of hard work, scraping and working alternately, over and over again in every direction, the skin can be made fairly soft. Keep at it until the desired flexibility is achieved. The more it is worked the softer it will become. After each scraping, add a little powdered alum and lay it away in a cool place. Another method of making skins pliable is to trample them in a tub of sawdust. This has a similar effect on skins as the commercial "drumming" process, in which the skins are tumbled endlessly about in a sawdust-filled, shelf-lined revolving drum.

MAKING BUCKSKIN

Home made buckskin can be produced from deer, elk, antelope, wolf, caribou and calf hides. The hide is soaked in clear water for a few hours, then fleshed thoroughly. The clean and fleshed hide is then put in the dehairing solution #5 overnight. Move it around occasionally the next day or two, and test it for loose hair. Always wear rubber gloves when working with this solution. When the hair slips off easily, scrape it away with a dull knife. (Another method of depilation or dehairing, making use of a lime water solution is described under Chrome Tanning.)

Now shave two cakes of naphtha soap in a little warm water, add a pint or so of Neats-foot oil and stir until a smooth mixture is obtained. Then add enough warm water to make three or four gallons of liquid tanning fluid.

After the solution has cooled, the dehaired and clean-scraped hide is immersed and allowed to remain submerged in the liquid for three days, during which the hide is occasionally stirred and turned. After the third day the hide is removed and squeezed almost dry and hung up until light spots begin to show. The skin should now be scraped with a dull tool or fleshing knife and "worked" back and forth over a fleshing beam until dry. Then return it to the tanning bath for another three days after which it is squeezed, scraped, hung up and worked, as before. The hide should come out smooth and soft.

If the buckskin has baggy spots or uneven edges from the beam work, wet the flesh side with a sponge, roll up until uniformly damp, then tack or lace it to a wall or frame, drawing it out smoothly and evenly. Release the buckskin when dry.

If desired, the buckskin may be smoked over a tiny wood fire to give a more pleasing odor and color. Drape the skin tent-like over a light framework of poles set up over the fire. It is important that the skin not be overheated as high heat will ruin it, but keep the fire very small and smothered with damp wood or shavings.

Indian Buckskin Making

American Indians made buckskin renowned for its toughness and durability. The following is a description of their methods.

The skins were fleshed and all membranous tissues removed. They were then soaked in water to swell and release impurities such as blood and dirt. The skins were grained with crude knives over a smooth log or fleshing beam. This way of removing the hair and the grain made buckskin of the highest class. Another way of preparing the skin for graining was to treat the skin with a weak lye solution for a few days, after which it was dehaired.

The skin having been grained, was then ready for the tanning process, which was done with the brains of the deer, or those from cattle, horses, etc. The brains were prepared by putting them into a bag of loosely-woven cloth and boiling them for about an hour in soft water. The water was then poured off into another container, and cooled until the hands could bear the heat. The bag was rubbed between the hands under this water until the brains were forced through the cloth. The buckskin was then kneaded and stretched in the liquor for about an hour, at intervals of ten minutes, and then was allowed to soak for several hours (overnight in cold weather), followed by more kneading and stretching. The skin was then partly dried, after which it was stretched every way until completely dry, then folded, wrapped, and put away for two weeks or more. The skin was now ready for the smoking process, which colored and retanned it into condition whereby it could be wetted repeatedly without becoming hard. Smoking was done with a slow fire of decayed wood, and was continued on both sides until the color was light yellow to yellowish-brown. The Indians stretched the skins over a barrel and conducted the smoke to it, or made a tent-like structure of branches. Smoke-houses used for curing meat were also used.

After being smoked, the skin was scoured in lukewarm water, rinsed, dried for a week, dipped in water for a few seconds, folded, and covered for a half-day, stretched and dried, and worked over a stake until soft and pliable.

TANNING SNAKESKINS

Results in snakeskin tanning depend largely on type and condition of the skin being processed. The shedding process with which snakes are endowed determines to a great extent the time when skins are in best condition for tannage, and results often leave much to be desired.

Skin snakes by cutting off the head, then making an incision on the centerline of the underside, from head end to tip of tail. Fasten to a post or rafter and peel the skin downward. Use a knife carefully where needed. Care must be taken to avoid pulling too hard, as some skins tear easily. Tanning can proceed immediately after skinning, while the skin is fresh. Scrape and clean off all adhering fat and bits of flesh.

If the skin is to be held for a while before tanning, it should be cured by soaking for two days in a brine composed of six pounds of salt mixed thoroughly with one pound of borax in a saturated water solution. Stir it occasionally, then hang it up to drain. When no longer dripping, roll it up loosely and smoothly, and dry it in a shady place. Do not freeze reptile skins!

When ready to tan, immerse the snakeskin for several days in a cold solution of 3 ounces of slaked lime to one gallon of warm water. At the end of this time, remove the scales by scraping lightly. Now soak the skin thoroughly in a weak boric-acid solution, prepared by adding one ounce boric acid to one gallon warm water and stirring until entirely dissolved. This will remove all traces of lime from the skin, which is very important. When this rinsing is completed, place the skin in a solution of 15-grams chrome alum and 1/4-pound salt, thoroughly mixed and dissolved in one gallon of water. Allow the skin to soak in this solution for three or four days. Now dissolve 5 grams of sodium carbonate in a half-pint of warm water and add this to the solution, drop by drop. Keep the skin in this solution for about a week, moving the skin about occasionally each day.

Then remove the skin, drain it, and soak it overnight in a solution consisting of one part sulphonated Neats-foot oil to three-parts water. Remove it the next day and drain the skin, then tack the skin on a board to dry. When thoroughly dry, work the skin over a fleshing beam until soft and pliable. Do not apply too great a pressure while working. The skin may then be pressed with an electric iron at low heat to flatten. A coat of flexible spray plastic or clear liquid celluloid may be applied to give a natural glassy sheen. Most snakeskins are sewn on a backing of contrasting colored felt with pinked edges.

CHROME TANNING FOR HEAVY LEATHER

Home tanning hides of large farm animals for sole, harness, lace, and other heavy-leather purposes, entails considerable effort and experience. It is practically impossible for the amateur to turn out leather of appearance and quality to compare with the commercial product, but it is quite possible for him to prepare serviceable leather suitable for many utility purposes in the home or on the farm.

A great many of the hides and skins taken from farm animals are removed by inexperienced persons and are seldom treated in an efficient manner with respect to skinning and curing. Most of the hides are in the "fallen" classification, that is, those coming from animals that have died by accident or natural causes, as well as those slaughtered for food purposes. This poor handling by farmers, ranchers, butchers and their helpers often results in an inferior grade of hides, hides with many imperfections, irregular pattern and trim, and hides yielding far less good-quality leather than would otherwise be the case.

Every effort should be made to do the hide skinning and curing in an efficient manner, keeping in mind constantly the value of the hide or skin as well as that of the meat. All hides and skins, large and small, should be evaluated and treated accordingly before any are discarded. Too many valuable hides and skins are discarded by thoughtless persons. Here again it should be remembered that animals suspected of disease should be destroyed immediately.

The average farm or range animal hide weighs about 50 pounds and it is for a single hide of about this weight that the following directions were prepared. Except that weaker tanning solutions should be used, these directions also apply for handling equivalent weight kips and the hides of smaller animals such as calf, deer, goats, etc., which may be used for thinner, lighter-weight leathers.

Skinning, Salting, Storing

It is important to have a good, sound, cleanly-skinned hide or skin for good leather production. It should be cleansed of all dirt and blood and be free of scores and knife cuts. All cuts should be clean and straight, and in skinning, the knife should be used as little as possible. The pattern of the hide should be equally distributed among shoulder, back, belly and butt sections. Of course, the hide must be free of meat, sinews, tail bone, horns, dew claws, etc.

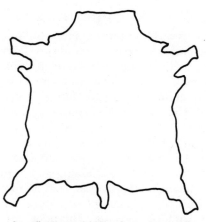

A well-trimmed hide of good pattern.

Ragged edges of shanks, head, tail, etc., are trimmed off before tanning.

Stretch the hide out on the floor, flesh-side up, and pour a pile of salt in the center. Use plenty of clean salt, at least a pound for each pound of hide. Spread the salt evenly toward the edges and be sure that every part is covered. The salt should be rubbed in with the hands.

If several hides are being salted, they may be stacked, hair-side down, upon each other. Single hides are best folded flesh-to-flesh and rolled up, after which they are placed on a sloping surface to drain. Hides and skins may be kept in such a green and salted condition for several months. Individual salted skins may be also safely kept for considerable periods of time by completely drying them out in a dry, airy place. Such hides must be protected from infestation by insects, however. Dried skins must never be folded or cracked and must be thoroughly soaked in clear water before handling.

Fleshing

Before the hide can be tanned, the following steps must be taken. Split the raw hide lengthwise down the center, dividing it into two "sides" for easier handling. The sides should then be thoroughly soaked and cleansed in a wooden 50-gallon barrel. After the hides have been washed and rinsed in several changes of clean water, allow them to soak for several hours. Cut two strong sticks, such as broomhandles, to a length that will fit crosswise into the barrel and fasten stout cords several feet long to each of the four ends. The sides can then be hung from the sticks and lowered into the water, flesh-side out. Fasten the cords to nails driven into the outside of the barrel. After the fresh, cleaned hides have soaked for about 18 hours, they should be removed and fleshed thoroughly. Use a large fleshing beam or the smooth-log substitute, as described under "Fleshing Beams."

It is important that this fleshing be thoroughly done and that every bit of flesh, fat and gristle be removed.

Dehairing

Dehairing the hide is the next step. To accomplish this, the hides are soaked in a limewater solution for ten to fourteen days. Prepare the solution in the barrel by adding 10 pounds of fresh, hydrated lime to five gallons of water. When the lime has been well mixed with water by stirring thoroughly, add enough clean water to nearly fill the barrel and stir again, using a large paddle. Now hang the two halves of the hide, hair-side out, over the stick supports and lower them into the limewater. Move the sides about occasionally and stir the solution several times each day. When the hair slips off very easily by hand rubbing, the hides may be removed and thrown over the beam where the hair is completely removed together with the top layer of thin skin in which it grows.

Turn the hides over on the beam and flesh them again, shaving thoroughly with a sharp currier's knife or substitute. The hides must now be cleaned of all traces of lime by rinsing and washing them several times in clear water, using several changes. After this, soak them again for 5 to 6 hours in clear water.

Now refill the barrel with clear, cool water and stir in five ounces of pure lactic acid. Vinegar may be used as a substitute for lactic acid, using five pints to the barrel, but the lactic acid is better.

Soak the sides in the lactic acid for 24 hours, moving them about occasionally, then hang them in clear water overnight. The hides are now ready to be tanned.

Tanning

Carefully weigh out the chemicals as listed in formula #9, and prepare the tanning solution as follows: Dissolve the 3½ lbs. of soda crystals and 6 lbs. of salt in 3 gallons of warm water in a wooden bucket. Then dissolve the 12 pounds of chrome alum in a tub containing 9 gallons of clear, cool water. Stir this mixture well. When the alum is completely dissolved add the soda-salt solution very slowly, stirring constantly. Now pour about 4 gallons of the combined solutions into the clean, empty barrel and add water until it is about two-thirds full. Hang the sides over the sticks and immerse in the tanning solution for three days, moving them about at short intervals. Avoid wrinkling or bunching-up of hides. When the three days have passed, add about four more gallons of the stock solution and allow the hides to tan for another three days, stirring and agitating them as before. On the sixth day add the remaining solution and allow the hides to tan three or four more days. The tanning process should now be complete. Cut a small sliver from each of the sides and if the edges are uniformly colored, tanning probably is complete. A reliable test can be made by boiling a small piece in water. If the piece curls up and becomes hard or rubbery, more soaking in the tanning fluid is indicated.

Lighter-weight or thinner hides will tan in a shorter time and require lesser amounts of chemicals and water. Prepare the solution exactly as given above, but in proportionately smaller amounts.

When tannage is completed, dispose of the tanning solution, wash the barrel thoroughly and give the hides a thorough rinsing in several changes of water. Then soak them in a solution of borax-water overnight. Make the solution by dissolving 2 lbs. of powdered borax to 40 gallons of water. (Use proportionately less water and borax for smaller hides.) The hides may now be removed from the borax solution and washed thoroughly through 5 or 6 changes of clean water before being hung up to drain.

Softening

The heavy leather should be oiled and "slicked" to smooth it out, free it of wrinkles, and to give it a better appearance. To do this, sprinkle the leather with warm water and when it is quite damp, apply a liberal coat of Neats-foot oil on both sides and slick the grain or hair-side vigorously with the slicking tool as described before. Then tack the hide to a wall or lace it on a sturdy frame, stretching the hide smoothly in either case. When dry, release and dampen the hide as before, slick and oil it again. Then apply a thick coat of warmed "dubbing" on the grain-side. Make the dubbing by melting equal parts of Neats-foot oil and tallow. If too thin, add more tallow until a smooth paste is formed. Hang the hide up again until dry, then slick off the excess dubbing. The leather can be rubbed with a soft cloth before using to remove surface oiliness.

The lighter-weight and thinner leather is oiled and finished in a slightly different manner. Let the hides dry slowly, but while still rather damp, give the grain (hair) side a good coating of warmed Neats-foot oil. Tack the hides on a wall or lace tightly in a frame and allow them to dry out completely. Then take them down again and dampen them with warm water and roll them up until uniformly limber. Slick the grain-side well in all directions, using plenty of pressure on the tool.

If softer and more supple leather is desired, the leather should be worked over the edge of a fleshing beam. If a regular fleshing beam is not available, the substitute "staking" method may be used as described under "Fleshing Beams." Repetition of this alternate oiling, slicking and "staking" of the leather will result in any degree of smoothness and pliability desired.

SKINS AND FURS — CARE AND CLEANING

Good furs deserve proper care. Here are the essentials on how to keep them clean and free from odor, and how to protect them against moth damage.

Deodorizing Furs and Hair Skins

Objectionable odors in certain furs and hair skins can be minimized or completely eliminated by the use of the following preparation:

Put about 4 pounds of finely-chipped bar soap into 2 gallons of water. Add about 4 pounds or slightly less, of sal soda and boil together until the soap and soda are dissolved. Then, while the solution is still hot, add 3/4 oz. of borax and 1/2 oz. of oil of sassafras. When the solution has cooled, it is ready for use. Rinse the skins well in this solution and when dried and cleaned, it will be found that the bad odor has disappeared. The solution can be used on skunk skins, etc., but these should be rinsed out before tanning. Proportionately smaller amounts of the ingredients can be used to make less solution.

Cleaning Furs

Dark furs can be cleaned with warm, hardwood sawdust such as cedar or mahogany. Lay the furs on a flat surface and work plenty of the warm sawdust into them. Rub vigorously and use plenty of the sawdust. After this, shake, beat and brush the fur until it is free of sawdust and comes out clean and fluffy. Bran, chaff, cornmeal, powdered borax, chalk, and dry plaster of paris may also be used as a cleaning agent. The use of the blower attachments of the common vacuum cleaner is recommended in these drying and fur-cleaning operations.

Soiled spots on furs may be cleaned by rubbing them with cubed magnesia or other solid commercial products now on the market. Some taxidermists wash the skins in warm soapy water or in borax-water solution, after which they dry the skins and rub them with plaster of paris. Professional cleaners dip fur garments, muffs, etc., into naphtha or gasoline until they are clean. After drying, a high gloss can be produced by special commercial preparations. Bread crust passed lightly with the direction of the nap will also produce a good gloss to the fur. Cleaning white furs or rugs may be accomplished by application of a paste of powdered chalk and water. Mix the two into a thin paste and rub thoroughly in the fur, then allowing the paste to dry completely. When dry, the material is thoroughly brushed out with a stiff brush and every particle removed. The furs should come out clean, soft and fluffy.

Mothproofing

Many a fine pelt has been ruined by damage caused by common house moth larvae. Fur skins and rugs can be protected from the ravages of these pests by hanging them in wind and sun periodically, after which they should be sprayed with any of the excellent commercial moth-repellent sprays now available.

Many taxidermists and naturalists recommend soaking tanned hides and furs in a borax-water solution made by adding enough powdered borax (about a quarter-pound to the gallon) to warm water to make a saturate solution. Let soak for about an hour, stirring and squeezing the skins occasionally. Then hang up in an airy place to dry. Skins treated thusly will be moth-proofed for a long period of time.

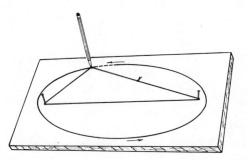

Making and finishing oval taxidermy display panels.

Use "C" grade, knot-free white pine or plywood, to make attractive but inexpensive oval display panels in any sizes. Lay out the desired oval shape with the use of two small nails, a piece of string that does not stretch, and a sharp pencil as shown. The length of the string loop and the relative distance between the nail settings will determine the size and shape of the panel.

The local planing mill will cut out the scribed oval with a band saw and turn an attractive edging on a shaper at little cost. Complete the panel by finishing with a wood antiquing kit obtainable at any paint store.

Attractive, antique-white finishes, for example, particularly adaptable for fish panels because of their fine contrasting effect, can be had by applying several coats of common interior white enamel undercoater. Brush the final coat in a wavy or curlique manner and allow it to dry. Then brush on a thin wash of burnt-umber oil pigment (or any other tube color) dissolved in paint-thinner spirits. Wipe this off almost immediately with a coarse cloth then, when dry, sand lightly with finish-grade garnet paper. A little hand-buffing with a soft cloth produces a lasting sheen.

Try to keep taxidermy display panels on the small side. A common tendency is to make them too large.

Glossary Of Terms

ALUM — A white astringent powder, used for shrinking skin fibers, hardening plaster, removing mucus from fish, etc., and in various tanning processes.

ANTLERS (AND HORNS) — Deer family *antlers* are composed of boney matter and are shed annually. A new heavier rack is grown each year, depending on feeding conditions. The *horns* of cattle, goats, buffalo, etc., on the other hand, are permanent and are composed of true horny tissue similar to the nails of humans, bills and claws of birds and animals. Both types are primarily fighting implements.

ARMATURE — A metal or wire framework used as a rigid support for building body forms or sculptures.

ARSENIC — (Arsenic trioxide.) A white crystalline powder, deadly poisonous. Used in insecticides, rodenticides and poisoning skins in taxidermy.

BEAMING — Working of tanned hides and skins over a tapered, rounded support to produce pliability and softness.

BORAX — A white crystalline salt, usually in powder form.

BUCKSKIN — A soft strong leather made from deerskins. Also used to describe similar types of leather collectively.

BUTT — The thick end of anything (such as the ear-butt), also the part of the hide or pelt that covered the animal's back and sides.

CARTILAGE — Tough, whitish animal-tissue or gristle. Solidifies when dry.

CARTILAGE KNIFE — A heavy scalpel with strong blade.

CASED — Skinned round, or having no lengthwise incision. Preferred manner of skinning most furbearers.

CAST — To form into a particular shape by pressing into a mold. A casting is an object so made.

CHROME (TANNING) — Mineral tanning with metallic salts as opposed to vegetable tannage.

CURING — Preserving hides by drying or salting, or both.

CURRYING — Introducing grease into leather and/or giving it desirable finish.

DEGREASING — Removal of grease by solvents or scraping.

DEHAIRING	The chemical removal of hair from hides, also a general term for removing hair.
DELIMING	Soaking skins or hides in a weak-acid or other solution to remove lime.
DEPILATION	The process following the soaking of hides, such as immersing them in an infusion of lime, which loosens the hair without injuring the skin.
DEPLETING (ACTION)	Any action which tends to loosen the hair and epidermis.
DERMIS	The true skin or that part from which leather is produced.
DEXTRIN	(Also dextrine.) A soluble gummy substance in powder form made from starch. Used in making adhesive pastes and hardening plaster, among other things.
DRUMMING	Operation of an apparatus in which hides and skins are tumbled as an aid in cleaning, tanning, fat-liquoring and stuffing.
DUBBIN (or DUBBING)	A mixture of oil and tallow for stuffing leather.
EAR LINERS	These are trimmed to fit the ear skins of animals and provide support and natural shape to the ears. Made of compressed paper, sheet lead, plastic, leather and similar materials.
EPIDERMIS	The external layer of the skin.
FAT-LIQUOR	An emulsion of soap and oil, a mild alkali, such as borax and oil, or a sulphonated oil, in which skins are worked after tanning and washing.
FLAT-SKINNED	Skinned open, or having a lengthwise incision in body and extremities allowing skin or hide to spread out.
FLESH-SIDE	The side of the skin or hide closest against the carcass.
FLESHING	Removal with a knife of portions of fat, flesh, tissue, gristle, etc. which adheres to a hide or skin.
FLUFFING	Producing a soft effect on the grain or flesh-side of leathers, also treatment of furs to produce soft, natural effect.
FORM	Artificial body, head, legs, etc. molded to exact natural shape, for use in taxidermy. Sometimes referred to as "manikin."
FORMALDEHYDE	(Formalin.) A colorless liquid, the commercial form is a forty percent solution. Properly diluted

with water, it becomes a powerful preservative and disinfectant. Formalin is rather unpleasant to handle because of its offensive odor and disagreeable tanning effect on the skin.

FURS — Skins tanned without removal of hair.

GRAIN — That portion of hide just under the epidermis. (Hair-side.)

GREEN (SKINS AND HIDES) — Fresh, untanned.

GREEN-SALTED — Fresh and salt-cured as applied to hides.

HAIR-SLIP — An effect produced by decomposition of grain, loss of hair due to putrefaction.

HIDES — General term for the skins of the larger animals.

KIPS — The skins of small beef cattle and calves. Also sometimes used to indicate sheep, lamb, pigs and other small animal skins.

LAMINATED — Composed of or arranged in thin sheets or layers to a desired thickness or strength.

LEATHERING — The process of turning raw hides and skins into leather.

LIMING — The soaking of hides in an infusion of lime, which process allows easy removal of hair and epidermis.

MAMMAL (Mammalia.) — Any of a group of vertebrate animals, the females of which have milk-secreting glands for feeding their offspring.

MANDIBLES — The two parts of a bird's beak.

MOLD — (Also "Mould.") A hollow form for giving a certain shape to something in a soft or liquid state.

NAPE — The back or lower side of the neck.

NEATS-FOOT OIL — A pale yellow fixed oil made by boiling the feet and shin bones of neat cattle, used in leather dressing.

PATTERN — The outline or shape of a hide after trimming.

PELT (or PELTRIES) — Animal skins in the fur or hair, or the aggregate thereof, usually referring to the furbearers.

PICKLING — Treating hides and skins with salt and sulphuric acid.

PLASTER OF PARIS — A heavy white powder, calcined gypsum, which when mixed with water forms a quick-setting paste or liquid for use in molding and casting.

POTTERS CLAY — A firm, plastic, fine-grained earth, chiefly

aluminum silicate, used for making ceramics, bricks, etc.

PREEN To clean and arrange feather tracts on bird skins.

PRESERVATIVE A substance applied to matter to prevent rotting or spoiling.

QUILLS Any of the large, stiff wing or tail feathers of a bird.

RAWHIDE Untanned and dehaired animal skin prepared by stretching and dry curing.

REGULATOR A straight, polished needle, six inches or more in length, useful in arranging the skin on a form, probing, setting eyes, and the like.

SCALP The head and neck skin of trophy animals. Also called the "cape."

SCALPEL A small, very keen-edged knife used in surgery, dissection, and other close cutting work.

SCORES Inadvertent cuts, notches or furrows made with skinning and fleshing tools causing flaws in hides or skins.

SEPARATOR A liquid or greasy material applied to surfaces of molds and casts to allow easy separation, prevent adhesion, etc.

SIDES Half hides, usually cut down centerline of the back for easier handling.

SKINS General term for the skins of the smaller animals.

SLICKING OUT Setting out and scraping to eliminate wrinkles and smooth the leather surface, also removing excess oil or water with a slicker.

SOAKING Washing skins or hides to remove dirt, blood and salt and soften for tanning.

STAKING Working a tanned skin or hide over a stake-beam to relax the fibers and promote softness and flexibility.

STUDY SKIN Natural skins of specimens prepared and preserved for identification, scientific study, collections, etc.

STUFFING The working-in of dubbin or oil.

STRUCK-THROUGH When the tanning agent has penetrated through the hide or skin.

SULPHONATED (OIL) An oil treated with sulphonic acid so that it contains sulphonic acids and hence is partly soluble in water.

TAN Any material which will prevent putrefaction of hide or skin substance.

TANNAGE A process of converting hide substance into leather.

TANNIC (ACID) | The active principle in many vegetable materials having the power of converting hide substance into leather.

TANNING | Treating prepared skins with an infusion containing tannic acid.

TAWING | Treating skins with compounds of aluminum.

TAXIDERMY | The art of preparing and preserving the skins of vertebrae and mounting them in a lifelike manner.

TISSUE | The cellular material of an organic body, as meat or flesh. Also, thin, unsized paper, nearly transparent.

TOUSING | Stretching, twisting and otherwise working a tanned skin to produce pliability and softness.

TOW | (Pronounced as "toe.") Soft, silky hemp or flax fiber used in taxidermy for wrapping bird and animal bodies, replacing muscles removed, filling, etc.

VISCERA | The internal organs of a body or carcass.

Index